Creation Calendar

Kalender der Schöpfung

Jewish Lunisolar Calendar 2024-2025

Jüdischer Lunisolarkalender 2024-2025

Elohically Publishings

by Shachar Haddad

Verlagslabel: Elohically Publishings

ISBN Softcover: 978-3-384-02042-0

Druck und Distribution im Auftrag der Autorin:
tredition GmbH, Heinz-Beusen-Stieg 5, 22926 Ahrensburg, Germany

Email: info@elohically-world.com
Internet: www.elohically-world.com

Calendar data by torahcalendar.com

Kalenderangaben von torahcalendar.com

Appreciation

This calendar's underlying data and information is due to the operators of the website *torahcalendar.com*, who through their hard work make the true calendar of the Bible and God [Elohim] available to all of us. Likewise, the website *bible-menorah.jimdofree.com* owes this calendar its intensive lunar knowledge.I print all this here, not because I want to falsely take all the credit, but in order to give all people, also those who are outside the Internet and cannot be online all the time, access to the true data and also to include it in the literature and in the circle of the personal organizers. Even though my hard work of painstakingly typing all these dates was equally immense, I imagine the compilation of the true calendar was certainly much more difficult and this present calendar hereby honors this hard work of these people who care about the true calendar. Errors on my part in typing may well be present, I apologize for that, it is human. Thank you very much.

Würdigung

Den diesem Kalender zugrundeliegenden Informationen sind den Betreibern der Internetseite torahcalendar.com *zu verdanken, die uns allen durch ihre harte Arbeit den wahren Kalender der Bibel und Gottes [Elohim] zugänglich machen. Ebenso die Internetseite* menora-bibel.jimdofree.com *verdankt dieser Kalender sein intensives Mondwissen.Ich drucke all dies hier ab, nicht, weil ich mich mit fremden Federn schmücken möchte, sondern, um allen Menschen, auch jenen, die sich außerhalb des Internets befinden und nicht andauernd online sein können, Zugang zu den wahren Daten zu geben und ihn auch in der Literatur und im Kreise der Terminplaner aufzunehmen. Auch wenn meine harte Arbeit des mühevollen Abtippens all dieser Daten ebenso immens war, so stelle ich mir vor, war das Zusammentragen des wahren Kalenders sicher schwieriger und dieser vorliegende Kalender würdigt hiermit diese harte Arbeit dieser Menschen, denen der wahre Kalender am Herzen liegt. Fehler meinerseits beim Abtippen können durchaus vorhanden sein, das bitte ich zu entschuldigen, das ist menschlich. Vielen Dank.*

Lunisolar calendar

The Jewish lunisolar calendar is contrary to the usual Jewish calendar the only correct calendar, because it was appointed by Elohim himself and can be read in the Torah (Leviticus 23) itself. This "Creation Calendar" or "God Calendar/Torah Calendar" is quite different from our well-known Gregorian Calendar, with which we all live.

In the 4th century after the Messiah, the Jewish Patriarch Hillel II invented the well-known "Jewish Calendar" used by the Jews since then and until today, so that the Jews living in the Babylonian exile at that time had a calendar by which they knew uniformly when which festivals were celebrated. At that time during the exile it may have been a very good thing, but when the exile was over and the Jews went back to Israel, they kept this false calendar and rejected the true calendar that Elohim gave them. This is the way it is to this day.

The calendar of Elohim is based on astronomical facts and is therefore very accurate, whereas the Jewish calendar is based only on statistical averages and is therefore inaccurate.
The calendar of Elohim described in the Tanakh is a purely astronomical calendar (lunisolar calendar: from Latin *luna* "moon" and *sol* "sun"), in which each month begins with the first visible crescent moon (the so-called "new light"). Due to the Diaspora (scattering of Jews across the earth), the Jews could not see the crescent moon over Jerusalem and therefore did not know when the new month began, making Hillel's Jewish calendar make sense, but after the exile Israel would have had to return to Elohim's calendar.

Elohim's lunisolar calendar ingeniously combines the celestial bodies (sun, moon, stars), nature (phenology), harvest seasons and time into a perfect whole. However, the Jewish calendar of Hillel is so inaccurate that it often deviates weeks or even months from the true dates and festivals!

- Elohim Yahuah has designated the month "Abib" (Nisan) as the first month of the year, whereas the Jewish/civil calendar designates the 7th month (Tishri) as the beginning of the year: Elohim tells us in the Book of Leviticus that the new year begins in spring, when life awakens anew, when the first ears of barley can be harvested, when it becomes brighter and brighter. The Jews celebrate New Year when everything becomes cold and dark in autumn, when the plant life ends and the harvest is already over.
- Elohim's calendar does not have a uniform number of days in a month. Months usually have either 29 or 30 days, but no more. How many days a month has is determined by the moon.
- Months do not have names, they are only numbered. Only the first month Abib (Aviv; Ex 34:18) has a name which means "barley ripening (ear ripening)". The names Nisan, Tishri, Elul, etc. were invented in the Babylonian exile, but should be avoided because they are not biblical.
- In Elohim's calendar, the ripeness of the early grain determines when the first month of the year begins; the Jews calculate the year and leap years mathematically, which results in inaccurate or even false dates regarding feasts and harvests, as well as

gathering times, so they gather at the wrong time and thus without Elohim. Their calculation takes no account of astronomy, which is crucial with Elohim. Thus the spring and autumn festivals are **always** near the equinox (March 19-21 & September 22-24), but in the Jewish calendar often a month after and thus much too late.

- So when the first month of a year began was always decided by the state of the early grain (barley), which had to be in the ripening stage "abib". If the barley was not yet ripe after checking, a leap month was inserted, so that there was always harmony between the calendar of Elohim and the harvest times.
- All festivals of Elohim are also harvest festivals in the three harvest seasons:
- 1) In spring (Pesach: barley),
- 2) In summer (Shavout: wheat),
- 3) In the fall (Yom Teruach/Kippur; Sukkot: grapes and fruits, later olives).
- The calendar of Elohim is a so-called "three-dimensional calendar" (3D), which works astronomically, agriculturally and geographically, thus combining nature and time into a perfect unity.
- Each day begins and ends at sunset, so that sunrise in the morning is always in the middle of an entire Jewish calendar day, which in turn ends at sunset. All biblical days do not begin at 24 o'clock, but already after sunset. It must always be paid attention to the new light, so the 1st day of all biblical months in the calendar of Elohim never begins in the eclipse (new moon), but always with the first appearance of the

moonlight (new light). So the phases of the moon determine the beginning of the month, but the position of the sun determines the beginning of the day (always after sunset). In this way, the Shabbat, the day of rest on the 7th day (from Friday sunset to Saturday sunset), also does not begin, as with the Jews, at 6 p.m., but always when the sun sets, which can be quite different summer and winter.

- Usually a year has 12 months, but since the lunar year is about 10.87 days shorter than the solar year, a leap year is inserted about every third year with an extra month. In this way, the calendar is brought into harmony with nature (phenology). The leap year causes that the feast days do not wander through the whole year, which is the case with the lunar calendar.
- The calendar of Elohim, of which we can read in the Bible, and the 7 Feast days of Elohim show a connection between the harvest times and the plan of Elohim for mankind:

I) Spring (Feast of Spring Fruits: Barley):

1) 1st month, Abib: 15th day: Passover 1, 1st day of Unleavened Bread (Passover 1).

2) 1st month, Abib: 21st day: Passover 7, 7th day of Unleavened Bread (Passover 7)

II) Summer (Feast of summer fruits: wheat):

3) The 50th day after the 1st Feast-Shabbat: Pentecost (Shavout/Festival of Weeks).

III) Autumn (Feast of late fruits: wine, fruit):

4) 7th month, 1st day: Trumpet Day (Yom Teruach).

5) 7th month, 10th day: Day of Atonement (Yom Kippur)

6) 7th month, 15th day: Feast of Tabernacles (Sukkot, 15th-21st day)
7) 7th month, 22nd day: Last Great Day (Shmini Azeret, 8th day of the gathering, conclusion)

Moonology

The calendar of Elohim provides that every new month begins with the appearance of the light in the darkness. The new month is the moment when after the new moon (invisible moon) for the first time the outline of the waxing moon is visible as a narrow crescent above the western horizon in the sky; this state of the moon is called new light.

Old light, on the other hand, is the waning crescent moon visible for the last time before a new moon.

Since the calendar of Elohim (Biblical calendar/Torah calendar) is a - as already mentioned - lunisolar calendar, it contains (like a lunar calendar) the 12 lunar months as calendar months, but for the approximation to the solar year (tropical year) on the average every three years a 13th lunar month is switched on (leap year).

The lunar year (354 days) is shorter than 365 days, the solar year longer. The average length of a tropical year is about 365 days, 5 hours, 48 minutes and 45 seconds - or 365.24219 days. The lunar phases have as their period the synodic month (lunation) with an average length of 29.53059 days (variations between about 29.27 and 29.83 days).

In the calendar, 6 months with 29 days and 6 months with 30 days are usually used alternately: Thus, the lunar calendar year is 354 days long (6x29+6x30). To adapt to the astronomically exact length of a lunar year with 354.3671 days (12 lunations), the calendrical lunar year is longer by one month approximately every three years. The beginning of each lunar month results from the observation, not a calculation!

Because the lunar year is about 10.87 days shorter than the solar year, these leap years are important to adapt the calendar to the state of nature, which is extremely important, because all festivals of Elohim are also harvest festivals. In this way, it has always been possible to ensure that the Passover always began with the barley harvest and the Pentecost always began with the wheat harvest.

To be able to see the first crescent moon (the new light) at all, there must be a minimum illuminated width, otherwise the light reflected by the sun does not reach the earth's surface. An illumination of 0.1% is not sufficient, there must be at least about 2%, even better 3-5%. Theoretically the new light over Jerusalem and thus also the beginning of the month could be determined mathematically over astronomical numerical values (e.g. from NASA), but it is not so simple, because still further criteria must be observed, like:
- Position of the sun,
- latitude,
- the season,
- the moon's orbit,

because the moon crescent must reach a certain minimum

height above the horizon, otherwise it becomes invisible by the extinction of the earth atmosphere.

Since the moon itself cannot produce light, but only reflects old light of the sun, there is basically no "new moon" at all, because it was never gone, it only looks like this, when the already past light of the sun does not shine on the moon any more. So it would be more logical, as in Switzerland, to speak of "full moon" (full light) and of "empty moon" (no light at all).

The term "new moon" describes the light constellation (moon phase) of the moon, which is not visible with the naked eye, when the moon is between earth and sun, i.e. in conjunction with the sun. During this constellation one sees from the earth only the night side of the moon. Only after approx. 35 hours the extreme right edge (left edge, if one observes from the southern half of the earth) is illuminated again by the direct sunlight.

This moon phase is the new light. The phase before it is called the new moon because the moon appears to "renew" itself during it. The last visible phase before the new moon is called "old light".

The 2nd new moon in a month and the 3rd new moon of a season with 4 new moons is called Black Moon.

The visibility of the new moon is called Rosh Chodesh in Hebrew.

Summary:

- Each month begins with the first visible crescent moon, called the new light.
- Full moon is always in the middle of the month, half moon in between (first/last quarter), because if no moonlight is visible, the month ends.
- Elohim's calendar connects the celestial stars with nature, defining time, i.e. if you see the state of the moon and nature (state of development of vegetation), you always know on which day of the year you are.
- The constellation of the stars gives you an additional information and makes it even easier, because it allows you to orient yourself not only in terms of time, but also spatially (geographically).

*[Source German text:*https://bible-menorah.jimdofree.com
English translation especially for this book by Shachar Haddad]

Lunisolarkalender

Der jüdische Lunisolarkalender ist entgegen des gewöhnlichen Jüdischen Kalenders der einzig richtige Kalender, denn er ist von Elohim selber eingesetzt worden und in der Tora (Levitikus 23) selber nachlesbar. Dieser „Kalender der Schöpfung (Creation Calendar)“ oder „Gotteskalender/Torakalender“ ist ganz anders als unser bekannter Gregorianischer Kalender, mit dem wir alle leben. Im 4.Jahrhundert nach dem Messias erfand der jüdische Patriarch Hillel II den von den Juden seitdem und bis zum heutigen Tage verwendeten bekannten „Jüdischen Kalender“, damit die damals im babylonischen Exil lebenden Juden einen Kalender hatten, an denen sie einheitlich wussten, wann welche Feste gefeiert wurden. Damals während des Exils mag es eine sehr gute Sache gewesen sein, doch als das Exil vorbei war und die Juden zurück nach Israel gingen, behielten sie diesen falschen Kalender bei und lehnten den wahren Kalender, den Elohim ihnen gab, ab. So ist das bis heute.
Der Kalender Elohims basiert auf astronomischen Gegebenheiten und ist dadurch sehr genau, wohingegen der Jüdische Kalender nur auf statistischen Mittelwerten beruht und dadurch ungenau ist.
Der im Tanach beschriebene Kalender Elohims ist ein rein astronomischer Kalender (Lunisolarkalender: von lat. *luna* „Mond“ und *sol* „Sonne“), bei dem jeder Monat mit der ersten sichtbaren Mondsichel (dem sogenannten „Neulicht“) beginnt. Durch die Diaspora (Zerstreuung der Juden über die Erde) konnten die Juden die Mondsichel über Jerusalem nicht sehen und wussten daher nicht, wann der neue Monat begann, wodurch der Jüdische Kalender von Hillel Sinn machte, nach

dem Exil jedoch hätte Israel zu Elohims Kalender zurückkehren müssen.
Elohims Lunisolarkalender verbindet genial die Himmelskörper (Sonne, Mond, Sterne), die Natur (Phänologie), die Erntezeiten und die Zeit zu einer perfekten Einheit. Der Jüdische Kalender Hillels ist jedoch so ungenau, dass er oft Wochen oder gar Monate von den wahren Daten und Festen abweicht!

- Elohim Yahuah hat den Monat „Abib“ (Nisan) als ersten Monat des Jahres bestimmt, der jüdische/bürgerliche Kalender sieht hingegen den 7.Monat (Tishri) als Jahresanfang vor: Elohim sagt uns im Buch Levitikus, dass das neue Jahre im Frühling beginnt, wenn das Leben neu erwacht, wenn die ersten Gerstenähren der Gerste geerntet werden können, wenn es immer heller wird. Die Juden feiern Neujahr, wenn im Herbst alles kalt und dunkel wird, wenn das Pflanzenleben endetund die Ernte schon vorbei ist.
- Der Kalender Elohims hat keine einheitliche Anzahl von Tagen im Monat. Monate haben meist entweder 29 oder 30 Tage, aber nicht mehr. Wie viele Tage ein Monat hat, bestimmt der Mond.
- Monate besitzen keine Namen, sie werden nur nummeriert. Nur der erste Monat Abib (Aviv; 2.Mose 34,18) hat einen Namen, der „Gerstenreife (Ährenreife)“ bedeutet. Die Namen Nisan, Tishri, Elul usw wurden im babylonischen Exil erfunden, sollten aber vermieden werden, da sie nicht biblisch sind.

- Der Reifegrad des Frühgetreides bestimmt im Kalender Elohims, wann der erste Monat des Jahres beginnt; die Juden berechnen das Jahr und die Schaltjahre mathematisch, was ungenaue oder sogar falsche Daten bezüglich Feste und Ernte ergibt sowie Versammlungszeiten, sodass sie sich zu falscher Zeit versammeln und dadurch ohne Elohim. Ihre Berechnung nimmt keine Rücksicht auf die Astronomie, die bei Elohim entscheidend ist. So liegen die Frühjahrs- und Herbst-Feste **immer** in der Nähe der Tag-und-Nacht-Gleiche (Equinox, 19.-21.März & 22.-24.September), aber im jüdischen Kalender oft einen Monat danach und damit viel zu spät.
- Wann also der erste Monat eines Jahres begann, entschied immer der Zustand des Frühgetreides (Gerste), das sich im Reifestadium „abib“ befinden musste. War die Gerste nach Überprüfung noch nicht reif, wurde ein Schaltmonat eingefügt, sodass stets Harmonie zwischen dem Kalender Elohims und den Erntezeiten herrschte.
- Alle Feste Elohims sind auch Erntefeste in den drei Erntezeiten:
  1) Im Frühling (Pesach: Gerste),
  2) Im Sommer (Shavout: Weizen),
  3) Im Herbst (Yom Teruach/Kippur; Sukkot: Trauben und Früchte, später Oliven).
- Der Kalender Elohims ist ein sogenannter „dreidimensionaler Kalender“, der astronomisch, landwirtschaftlich und geografisch funktioniert und

dadurch die Natur und die Zeit zu einer perfekten Einheit verbindet.

- Jeder Tag beginnt und endet bei Sonnenuntergang, sodass der Sonnenaufgang am Morgen stets in der Mitte eines ganzen jüdischen Kalendertages liegt, der wiederum bei Sonnenuntergang beendet wird. Alle biblischen Tage beginnen nicht um 24 Uhr, sondern bereits nach Sonnenuntergang. Es muss immer auf das Neulicht geachtet werden, so beginnt der 1.Tag aller biblischen Monate im Kalender Elohims niemals in der Finsternis (Neumond), sondern immer mit dem ersten Erscheinen des Mondlichts (Neulicht). Die Mondphasen bestimmen also den Monatsanfang, aber der Sonnenstand bestimmt den Tagesanfang (jeweils nach Sonnenuntergang). Auf diese Weise beginnt der Shabbat, der Ruhetag am 7.Tag (von Freitag Sonnenuntergang bis Samstag Sonnenuntergang), auch nicht, wie bei den Juden, um 18 Uhr, sondern immer dann, wenn die Sonne untergeht, was Sommer wie Winter ganz unterschiedlich sein kann.
- Für gewöhnlich hat ein Jahr 12 Monate, doch da das Mondjahr etwa 10,87 Tage kürzer ist als das Sonnenjahr, wird etwa jedes dritte Jahr ein Schaltjahr eingefügt mit einem zusätzlichen Monat. Auf diese Weise wird der Kalender mit der Natur in Einklang gebracht (Phänologie). Das Schaltjahr bewirkt, dass die Festtage nicht durch das ganze Jahr wandern, was beim Mondkalender der Fall ist.
- Der Kalender Elohims, von dem wir in der Bibel nachlesen können, und die 7 Festtage Elohims zeigen einen Zusammenhang zwischen den Erntezeiten und

dem Plan Elohims für die Menschheit:

I) Frühjahr (Fest der Frühlingsfrüchte: Gerste):

1) 1.Monat, Abib: 15.Tag: Passah 1, 1.Tag der Ungesäuerten Brote (Pessach 1)

2) 1.Monat, Abib: 21.Tag: Passh 7, 7.Tag der Ungesäuerten Brote (Pessach 7)

II) Sommer (Fest der Sommerfrüchte: Weizen):

3) Der 50.Tag nach dem 1.Fest-Shabbat: Pfingsten (Shavout/Wochenfest)

III) Herbst (Fest der Spätfrüchte: Wein, Obst):

4) 7.Monat, 1.Tag: Posaunentag (Yom Teruach)

5) 7.Monat, 10.Tag: Versöhnungstag (Yom Kippur)

6) 7.Monat, 15.Tag: Laubhüttenfest (Sukkot, 15.-21.Tag)

7) 7.Monat, 22.Tag: Letzter Großer Tag (Shmini Azeret, 8.Tag der Versammlung, Abschluss)

Mondkunde

Der Kalender Elohims sieht vor, dass jeder neue Monat mitdem Erscheinen des Lichts in der Finsternis beginnt. Als neuer Monat wird jener Moment bezeichnet, wenn nach Neumond (unsichtbarer Mond) zum esten Mal die Kontur des zunehmenden Mondes als schmale Sichel über dem westlichen Horizont am Himmel zu sehen ist; dieser Zustand des Mondes nennt sich Neulicht.

Als Altlicht hingegen wird die zum letzten Mal vor einem

Neumond sichtbare abnehmende Mondsichel bezeichnet.
Da der Kalender Elohims (Biblischer Kalender/Torahkalender) ein – wie bereits erwähnt – Lunisolarkalender ist, enthält er zwar (wie ein Lunarkalender) die 12 Mond-Monate als Kalender-Monate, doch zur Annäherung an das Sonnenjahr (Tropisches Jahr) wird durchschnittlich alle drei Jahre ein 13.Mond-Monat eingeschaltet (Schaltjahr).
Das Mondjahr (354 Tage) ist kürzer als 365 Tage, das Sonnenjahr länger. Die durchschnittliche Länge eines tropischen Jahres beträgt etwa 365 Tage, 5 Stunden, 48 Minuten und 45 Sekunden – oder 365,24219 Tage. Die Mondphasen haben als Periode den synodischen Monat (Lunation) mit einer durchschnittlichen Länge von 29,53059 Tagen (Schwankungen zwischen etwa 29,27 und 29,83 Tagen).
Im Kalender werden in der Regel 6 Monate mit 29 Tagen und 6 Monate mit 30 Tagen im Wechsel verwendet: Damit ist das kalendarische Mondjahr 354 Tage lang (6x29+6x30). Zur Anpassung an die astronomisch genaue Länge eines Mondjahres mit 354,3671 Tagen (12 Lunationen) ist das kalendarische Mondjahr etwa alle drei Jahre um einen Monat länger. Der Beginn jedes Mondmonats ergibt sich aus der Beobachtung, nicht einer Berechnung!
Weil das Mondjahr etwa 10,87 Tage kürzer ist als das Sonnenjahr, sind diese Schaltjahre wichtig, um den Kalender an den Zustand der Natur anzupassen, was extrem wichtig ist, da alle Feste Elohims auch Erntedank-Feste sind. Auf diese Weise konnte immer sichergestellt werden, dass beim Passahfest immer mit der Gerstenernte und beim Pfingstfest immer mit der Weizenernte begonnen werden konnte.
Um die erste Mondsichel (das Neulicht) überhaupt sehen zu

können, muss eine beleuchtete Mindestbreite vorhanden sein, da ansonsten das von der Sonne reflektierte Licht auf der Erdoberfläche nicht ankommt. Eine Beleuchtung von 0,1% reicht nicht aus, es müssen mindestens ca. 2% vorhanden sein, noch besser sind 3-5%. Theoretisch könnte über astronomische Zahlenwerte (z.B. von der NASA) das Neulicht über Jerusalem und somit auch der Monatsanfang rechnerisch bestimmt werden, doch so einfach ist es nicht, denn es müssen noch weitere Kriterien beobachtet werden, wie:

- Sonnenstand,
- geografische Breite,
- die Jahreszeit,
- die Mondbahn,

denn die Mondsichel muss eine gewisse Mindesthöhe über dem Horizont erreichen, andernfalls wird sie durch die Extinktion der Erdatmosphäre unsichtbar.

Da der Mond selber kein Licht erzeugen kann, sondern nur altes Licht der Sonne reflektiert, gibt es im Grunde gar keinen „Neumond", denn er war nie weg, es sieht nur so aus, wenn das bereits vergangene Licht der Sonne den Mond nicht mehr bescheint. So wäre es logischer, wie in der Schweiz, von „Vollmond" (volles Licht) und von „Leermond" (gar kein Licht) zu sprechen.

Der Begriff „Neumond" bezeichnet die mit bloßem Auge nicht erkennbare Lichtgestalt (Mondphase) des Mondes, wenn sich dieser zwischen Erde und Sonne, d.h. in Konjunktion mit der Sonne befindet. Während dieser Konstellation sieht man von der Erde aus nur die Nachtseite des Mondes. Erst nach ca. 35 Stunden wird der äußerste rechte Rand (linke Rand, wenn man

von der südlichen Erdhälfte aus beobachtet) wieder vom direkten Sonnenlicht erhellt.
Diese Mondphase ist das Neulicht. Die Phase vorher heißt Neumond, weil sich der Mond in ihr scheinbar „erneuert“. Die letzte sichtbare Phase vor dem Neumond heißt „Altlicht“.
Der 2.Neumond in einem Monat und der 3.Neumond einer Jahreszeit mit 4 Neumonden wird als Black Moon bezeichnet.
Das Sichtbarwerden des Neumondes wird auf Hebräisch Rosh Chodesh genannt.

Zusammenfassung:

- Jeder Monat beginnt mit der ersten sichtbaren Mondsichel, dem sogenannten Neulicht.
- Vollmond ist immer in der Mitte des Monats, Halbmond dazwischen (Erstes/Letztes Viertel), denn wenn kein Mondlicht zu sehen ist, dann geht der Monat zuende.
- Der Kalender Elohims verbindet die Himmelsgestirne mit der Natur, womit die Zeit definiert wird, d.h. wenn man den Zustand des Mondes und der Natur (Entwicklungsstand der Vegetation) sieht, weiß man immer, an welchem Tag des Jahres man sich befindet.
- Die Konstellation der Sterne gibt einem eine zusätzliche Information und macht es noch leichter, weil man sich dadurch nicht nur in Bezug auf die Zeit, sondern auch räumlich (geografisch) orientieren kann.

(Quelle: https://menora-bibel.jimdofree.com)

How to use this Calendar

This special Jewish lunisolar calendar or also called "calendar of the creation" is divided differently than every usual Gregorian calendar which one knows and uses so. Neither the days of the week nor the months have a name in Hebrew. They were indeed divided in the Babylonian exile of the Jews into "Nisan" or "Tammuz" and so on, but only the first month in the Hebrew year, which begins in spring (New Year), has a name: Abib. The calendar before you starts with 1 in the Hebrew month and goes on to 29 or 30, because each month is based on the moon. No Hebrew month lasts more than 30 days. The large numbers you see on each day are the days of the Hebrew month, so please don't confuse them with the month numbers of the Gregorian calendar, which you can read much smaller above (see chart next page).

The calendar used by the Jews is called the Jewish calendar, often even the Biblical calendar, but this calendar is wrong. *"In some years there are differences between God's calendar mentioned in the Bible and the Jewish calendar of today. The latter was created centuries after the destruction of Jerusalem (70 AD), because the Jews, scattered all over the world, could no longer see the new light of the moon (important for determining the beginning of the month) over Jerusalem, but all should rest on the same feast days. The binding establishment of the Jewish calendar was made in 359 A.D. by Patriarch Hillel II. Consequently, the Jewish calendar is not the same calendar that the Israelites had at the time of Yahushua the Messiah.. During the almost 2,000-year period of the dispersion of the people of Israel among all peoples, Hillel's calendar made sense; but at the latest since the return of the Jews to Israel (1948) and Jerusalem (1967), there is nothing to stop us from using God's original calendar again, especially since the laws of nature (astronomy) about moonlight have never changed."* (source: https://bible-menorah.jimdofree.com/english/calendar-and-feasts/)

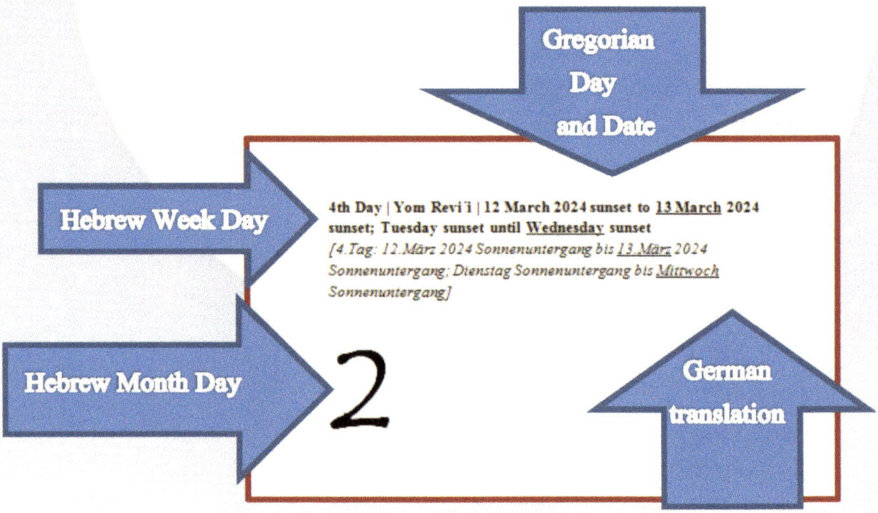

Wie man diesen Kalender benutzt

Dieser besondere jüdische Lunisolarkalender oder auch »Kalender der Schöpfung« genannt, wird anders eingeteilt als jeder gewöhnliche gregorianische Kalender, den man so kennt und nutzt. Weder die Wochentage noch die Monate haben im Hebräischen einen Namen. Sie wurden zwar im babylonischen Exil der Juden in »Nisan« oder »Tammuz« und so weiter eingeteilt, doch nur der erste Monat im hebräischen Jahr, der im Frühling beginnt (Neujahr), besitzt einen Namen: Abib.
Der vor dir liegende Kalender fängt im hebräischen Monat mit 1 an und geht weiter bis 29 oder 30, denn jeder Monat richtet sich nach dem Mond. Kein hebräischer Monat dauert mehr als 30 Tage. Die großen Ziffern, die du an jedem Tag siehst, sind die Tage des hebräischen Monats, also bitte nicht verwechseln mit den Monatszahlen des gregorianischen Kalenders, welche du weiter oben viel kleiner ablesen kannst (siehe Grafik nächste Seite).
Der von den Juden verwendete Kalender wird als Jüdischer Kalender bezeichnet, oft sogar als Biblischer Kalender, doch dieser Kalender ist falsch. *»In einigen Jahren gibt es Unterschiede zwischen dem in der Bibel erwähnten **Kalender Gottes** und dem heutigen **jüdischen Kalender**. Letzterer wurde erst Jahrhunderte nach der Zerstörung Jerusalems (70 n. Chr.) erstellt, da die weltweit verstreuten Juden nicht mehr das Neulicht des Mondes (wichtig zur Bestimmung des Monatsanfangs) über Jerusalem sehen konnten, aber alle zu den gleichen Festtagen ruhen sollten. Die verbindliche*

Festlegung des ***jüdischen Kalenders*** *erfolgte im Jahr* ***359 n. Chr.*** *durch den* ***Patriarchen Hillel II****. Es handelt sich beim jüdischen Kalender folglich nicht um den gleichen Kalender, den die Israeliten zur Zeit Yahushua des Messias hatten. Während der fast 2.000-jährigen Zeit der Zerstreuung des Volkes Israel unter allen Völkern machte der* ***Kalender von Hillel*** *Sinn; aber spätestens seit der Rückkehr der Juden nach Israel (1948) und Jerusalem (1967) spricht nichts dagegen, den ursprünglichen* ***Kalender Gottes*** *wieder zu verwenden, zumal sich die Naturgesetze (Astronomie) über das Mondlicht niemals verändert haben« (Quelle: menora-bibel.jimdofree.com).*

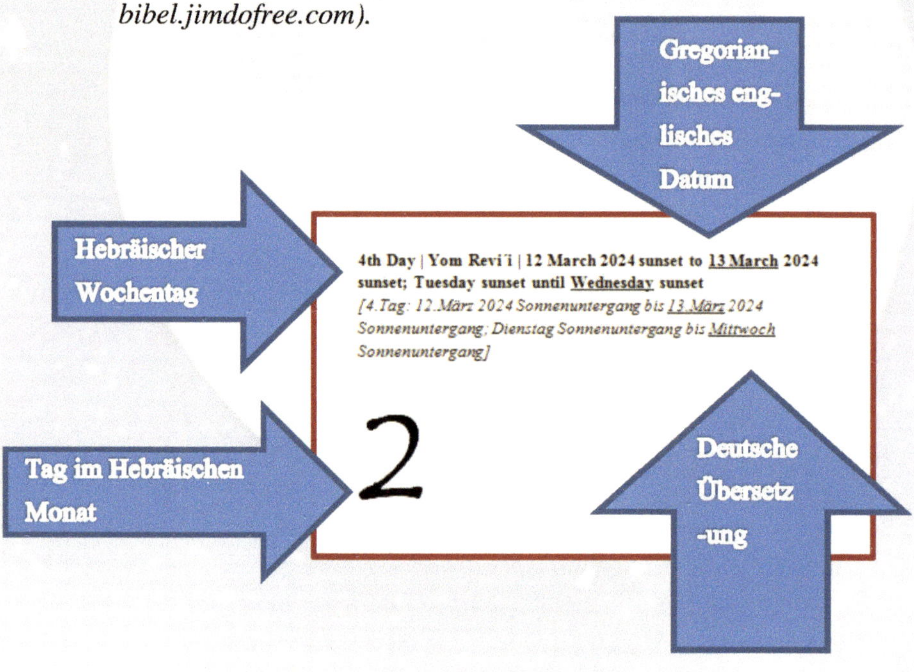

Moon Phases | *Mondphasen*

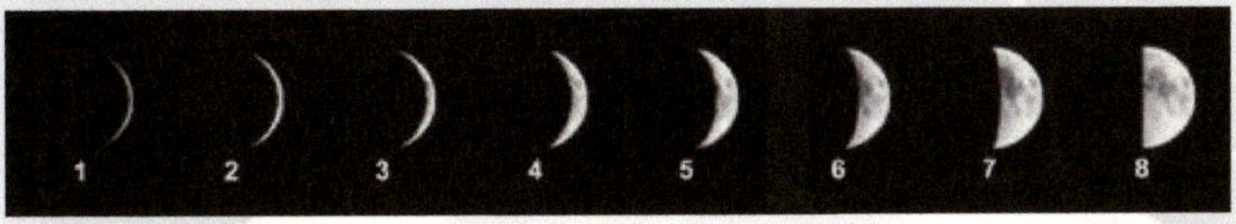

1: New light (first visible crescent moon) | *Neulicht (erste sichtbare Mondsichel)/Neumond*
7/8: First Quater (Halfmoon) | *Erstes Viertel (Halbmond)*

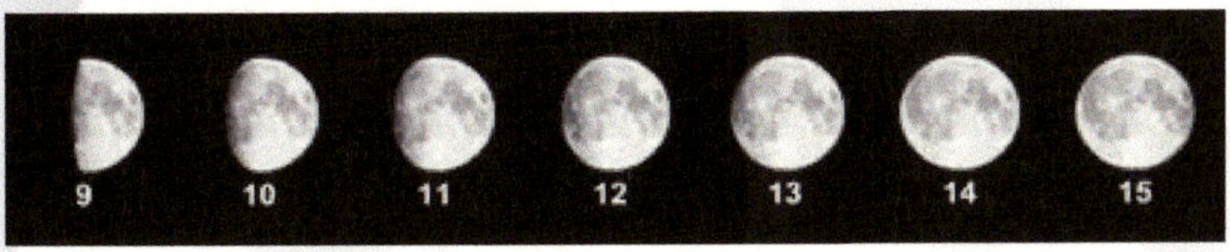

14/15: Full Moon (always in the middle of the Hebrew month) | *Vollmond (immer Mitte des hebräischen Monats)*

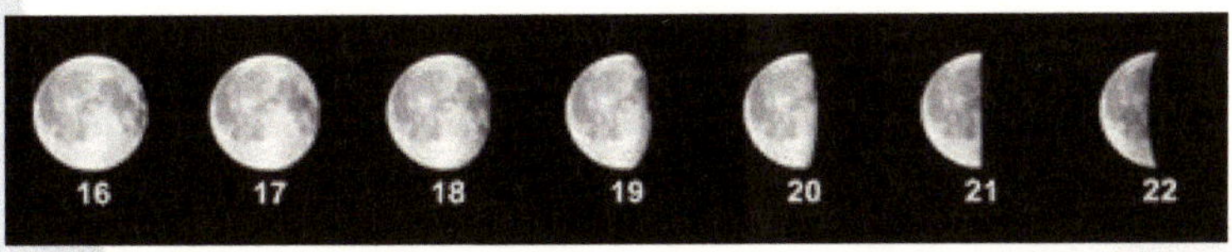

21: Last Quater (Halfmoon)| *Letztes Viertel (Halbmond)*

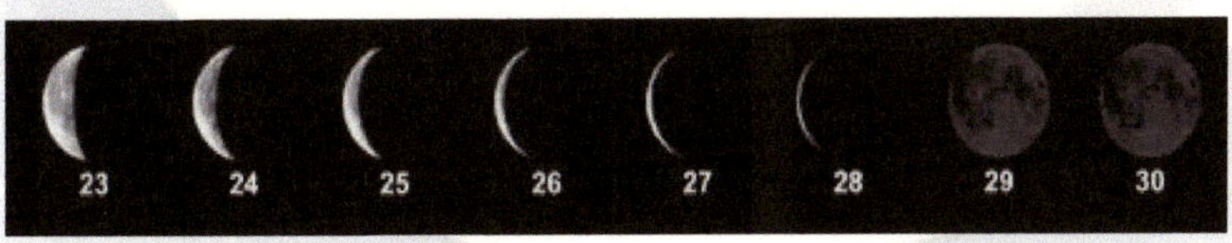

28: Old light (waning crescent visible) | *Altlicht (sichtbare abnehmende Mondsichel)*
29/30: New Moon | *Neumond*

(Source chart: https://bible-menorah.jimdofree.com/english/calendar-and-feasts/)
(Quelle Grafik: menora-bibel.jimdofree.com)

Month 10 | *Monat 10*

to start with gregorian Calender here for using it on 1st January in our calendar|

Hier starten wir mit dem Gregorianischen Kalender, um mit dem 1.Januar in unserem Kalender beginnen zu können

---

2nd Day | Yom Sheni | 31 December 2023 sunset to <u>1 January</u> 2024 sunset; Sunday sunset until <u>Monday</u> sunset

[2.Tag. 31.Dezember 2023 Sonnenuntergang bis 1.Januar 2024 Sonnenuntergang; Sonntag Sonnenuntergang bis Montag Sonnenuntergang]

18

3rd Day | Yom Shlishi | 1 January 2024 sunset to 2 January 2024 sunset; Monday sunset until Tuesday sunset

[3.Tag: 1.Januar 2024 Sonnenuntergang bis 2.Januar 2024 Sonnenuntergang; Montag Sonnenuntergang bis Dienstag Sonnenuntergang]

19

4th Day | Yom Revi´i | 2 January 2024 sunset to 3 January 2024 sunset; Tuesday sunset until Wednesday sunset
[4.Tag: 2.Januar 2024 Sonnenuntergang bis 3.Januar 2024 Sonnenuntergang; Dienstag Sonnenuntergang bis Mittwoch Sonnenuntergang]

20

5th Day | Yom Hamishi | 3 January 2025 sunset to 4 January 2024 sunset; Wednesday sunset until Thursday sunset

[5.Tag: 3.Januar 2024 Sonnenuntergang bis 4.Januar 2024 Sonnenuntergang; Mittwoch Sonnenuntergang bis Donnerstag Sonnenuntergang]

21

6th Day | Yom Shishi | 4 January 2024 sunset to 5 January 2024 sunset; Thursday sunset until Friday sunset

[6.Tag: 4.Januar 2024 Sonnenuntergang bis 5.Januar 2024 Sonnenuntergang; Donnerstag Sonnenuntergang bis Freitag Sonnenuntergang]

22

7th Day | Shabbat | 5 Janaury 2024 sunset to 6 January 2024 sunset; Friday sunset until Saturday sunset

[7.Tag: 5.Januar 2024 Sonnenuntergang bis 6.Januar 2024 Sonnenuntergang; Freitag Sonnenuntergang bis Samstag Sonnenuntergang]

23

1st Day | Yom Rishon | 6 January 2024 sunset to 7 January 2024 sunset; Saturday sunset until Sunday sunset

[1.Tag: 6.Januar 2024 Sonnenuntergang bis 7.Januar 2024 Sonnenuntergang; Samstag Sonnenuntergang bis Sonntag Sonnenuntergang]

24

2nd Day | Yom Sheni | 7 January 2024 sunset to 8 January 2024 sunset; Sunday sunset until Monday sunset

[2.Tag: 7.Januar 2024 Sonnenuntergang bis 8.Januar 2024 Sonnenuntergang; Sonntag Sonnenuntergang bis Montag Sonnenuntergang]

25

3rd Day | Yom Shlishi | 8 January 2024 sunset to 9 January 2024 sunset; Monday sunset until Tuesday sunset

[3.Tag: 8.Januar 2024 Sonnenuntergang bis 9.Januar 2024 Sonnenuntergang; Montag Sonnenuntergang bis Dienstag Sonnenuntergang]

26

4th Day | Yom Revi´i | 9 January 2024 sunset to 10 January 2024 sunset; Tuesday sunset until Wednesday sunset: Old Light

[4.Tag: 9.Januar 2024 Sonnenuntergang bis 10.Januar 2024 Sonnenuntergang; Dienstag Sonnenuntergang bis Mittwoch Sonnenuntergang: Altlicht]

27

5th Day | Yom Hamishi | 10 January 2024 sunset to 11 January 2024 sunset; Wednesday sunset until Thursday sunset: New Moon

[5.Tag: 10.Januar 2024 Sonnenuntergang bis 11.Januar 2024 Sonnenuntergang; Mittwoch Sonnenuntergang bis Donnerstag Sonnenuntergang: Neumond]

28

6th Day | Yom Shishi | Rosh Chodesh: First Crescent Moon, sunset 12 Jan: New Light | 11 January 2024 sunset to 12 January 2024 sunset; Thursday sunset until Friday sunset

[6.Tag: Rosh Chodesh: Erste sichtbare Mondsichel, Sonnenuntergang 12.Jan: Neulicht | 11.Januar 2024 Sonnenuntergang bis 12.Januar 2024 Sonnenuntergang; Donnerstag Sonnenuntergang bis Freitag Sonnenuntergang]

29 

Month 11 | *Monat 11*

7th Day | Shabbat/Shabbat Rosh Chodesh | Rosh Chodesh: New Light | 12 January 2024 sunset to 13 January 2024 sunset; Friday sunset until Saturday sunset

[7.Tag: Rosh Chodesh: Neulicht | 12.Januar 2024 Sonnenuntergang bis 13.Januar 2024 Sonnenuntergang; Freitag Sonnenuntergang bis Samstag Sonnenuntergang]

1 

1st Day | Yom Rishon | Death of Herod (Matthew 2:19); Megillat Ta´anit | 13 January 2024 sunset to 14 January sunset; Saturday sunset until Sunday sunset

[1.Tag: Tod von Herodes (Matthäus 2,19); Megillat Ta´anit | 13.Januar 2024 Sonnenuntergang bis 14.Januar 2024 Sonnenuntergang; Samstag Sonnenuntergang bis Sonntag Sonnenuntergang]

2

4th Day | Yom Revi´i | 16 January 2024 sunset to 17 January 2024 sunset; Tuesday sunset until Wednesday sunset

[4.Tag: 16.Januar 2024 Sonnenuntergang bis 17.Januar 2024 Sonnenuntergang; Dienstag Sonnenuntergang bis Mittwoch Sonnenuntergang]

5

5th Day | Yom Hamishi | 17 January 2024 sunset to 18 January 2024 sunset; Wednesday sunset to Thursday sunset

[5.Tag: 17.Januar 2024 Sonnenuntergang bis 18.Januar 2024 Sonnenuntergang; Mittwoch Sonnenuntergang bis Donnerstag Sonnenuntergang]

6

6th Day | Yom Shishi | 18 January 2024 sunset to 19 January 2024 sunset; Thursday sunset until Friday sunset

[6.Tag: 18.Januar 2024 Sonnenuntergang bis 19.Januar 2024 Sonnenuntergang; Donnerstag Sonnenuntergang bis Freitag Sonnenuntergang]

7

7th Day | Shabbat | 19 January 2024 sunset to 20 January 2024 sunset; Friday sunset until Saturday sunset

[7.Tag: 19.Januar 2024 Sonnenuntergang bis 20.Januar 2024 Sonnenuntergang; Freitag Sonnenuntergang bis Samstsg Sonnenuntergang]

8

1st Day | Yom Rishon | 20 January 2024 sunset to 21 January 2024 sunset; Saturday sunset until Sunday sunset

[1.Tag: 20.Januar 2024 Sonnenuntergang bis 21.Januar 2024 Sonnenuntergang; Samstag Sonnenuntergang bis Sonntag Sonnenuntergang]

9

2nd Day | Yom Sheni | 21 January 2024 sunset to 22 January 2024 sunset; Sunday sunset until Monday sunset

[2.Tag: 21.Januar 2024 Sonnenuntergang bis 22.Januar 2024 Sonnenuntergang; Sonntag Sonnenuntergang bis Montag Sonnenuntergang]

10

3rd Day | Yom Shlishi | 22 January 2024 sunset to 23 January 2024 sunset; Monday sunset until Tuesday sunset

[3.Tag: 22.Januar 2024 Sonnenuntergang bis 23.Januar 2024 Sonnenuntergang; Montag Sonnenuntergang bis Dienstag Sonnenuntergang]

11

4th Day | Yom Revi´i | 23 January 2024 sunset to 24 January 2024 sunset; Tuesday sunset until Wednesday sunset
[4.Tag: 23.Januar 2024 Sonnenuntergang bis 24.Januar 2024 Sonnenuntergang; Dienstag Sonnenuntergang bis Mittwoch Sonnenuntzergang]

12

5th Day | Yom Hamishi | 24 January 2024 sunset to 25 January 2024 sunset; Wednesday sunset until Thursday sunset

[5.Tag: 24.Januar 2024 Sonnenuntergang bis 25.Januar 2024 Sonnenuntergang; Mittwoch Sonnenuntergang bis Donnerstag Sonnenuntergang]

13

6th Day | Yom Shishi | 25 January 2024 sunset to 26 January 2024 sunset; Thursday sunset until Friday sunset: Full Moon

[6.Tag: 25.Januar 2024 Sonnenuntergang bis 26.Januar 2024 Sonnenuntergang; Donnerstag Sonnenuntergang bis Freitag Sonnenuntergang: Vollmond]

14

7th Day | Shabbat | Tu B´Shvat: New Year for Trees (Rosh Hashanah 14A & B) | 26 January 2024 sunset to 27 January 2024 sunset; Friday sunset until Saturday sunset

[7.Tag: Tu B´Shvat: Neujahr der Bäume (Rosh Hashana 14 A & B) | 26.Januar 2024 Sonnenuntergang bis 27.Januar 2024 Sonnenuntergang; Freitag Sonnenuntergang bis Samstag Sonnenuntergang]

15

1st Day | Yom Rishon | 27 January 2024 sunset to 28 January 2024 sunset; Saturday sunset until Sunday sunset

[1.Tag: 27.Januar 2024 Sonnenuntergang bis 28.Januar 2024 Sonnenuntergang; Samstag Sonnenuntergang bis Sonntag Sonnenuntergang]

16

2nd Day | Yom Sheni | 28 January 2024 sunset to 29 January 2024 sunset; Sunday sunset until Monday sunset

[2.Tag: 28.Januar 2024 Sonnenuntergang bis 29.Januar 2024 Sonnenuntergang; Sonntag Sonnenuntergang bis Montag Sonnenuntergang]

17

3rd Day | Yom Shlishi | 29 January 2024 sunset to 30 January 2024 sunset; Monday sunset until Tuesday sunset

[3.Tag: 29.Januar 2024 Sonnenuntergang bis 30 Januar 2024 Sonnenuntergang; Montag Sonnenuntergang bis Dienstag Sonnenuntergang]

18

4th Day | Yom Revi´i | 30 January 2024 sunset to 31 January 2024 sunset; Tuesday sunset until Wednesday sunset

[4.Tag: 30.Januar 2024 Sonnenuntergang bis 31.Januar 2024 Sonnenuntergang; Dienstag Sonnenuntergang bis Mittwoch Sonnenuntergang]

19

5th Day | Yom Hamishi | 31 January 2024 sunset to <u>1 February</u> 2024 sunset; Wednesday sunset until <u>Thursday</u> sunset

[5.Tag: 31.Januar 2024 Sonnenuntergang bis 1.Februar 2024 Sonnenuntergang; Mittwoch Sonnenuntergang bis Dienstag Sonnenuntergang]

20

6th Day | Yom Shishi | 1 February 2024 sunset to 2 February 2024 sunset; Thursday sunset until Friday sunset

[6.Tag: 1.Februar 2024 Sonnenuntergang bis 2.Februar 2024 Sonnenuntergang; Donnerstag Sonnenuntergang bis Freitag Sonnenuntergang]

21

7th Day | Shabbat | 2 February 2024 sunset to 3 February 2024 sunset; Friday sunset until Saturday sunset

[7.Tag: 2.Februar 2024 Sonnenuntergang bis 3.Februar 2024 Sonnenuntergang; Freitag Sonnenuntergang bis Samstag Sonnenuntergang]

22

1st Day | Yom Rishon | 3 February 2024 sunset to 4 February 2024 sunset; Saturday sunset until Sunday sunset

[1.Tag: 3.Februar 2024 Sonnenuntergang bis 4.Februar 2024 Sonnenuntergang; Samstag Sonnenuntergang bis Sonntag Sonnenuntergang]

23

2nd Day | Yom Sheni | 4 February 2024 sunset to 5 February 2024 sunset; Sunday sunset until Monday sunset

[2.Tag: 4.Februar 2024 Sonnenuntergang bis 5.Februar 2024 Sonnenuntergang; Sonntag Sonnenuntergang bis Montag Sonnenuntergang]

24

3rd Day | Yom Shlishi | 5 February 2024 sunset to 6 February 2024 sunset; Monday sunset until Tuesday sunset

[3.Tag: 5.Februar 2024 Sonnenuntergang bis 6.Februar 2024 Sonnenuntergang; Montag Sonnenuntergang bis Dienstag Sonnenuntergang]

25

4th Day | Yom Revi´i | 6 February 2024 sunset to 7 February 2024 sunset; Tuesday sunset until Wednesday sunset

[4.Tag: 6.Februar 2024 Sonnenuntergang bis 7.Februar 2024 Sonnenuntergang; Dienstag Sonnenuntergang bis Mittwoch Sonnenuntergang]

26

5th Day | Yom Hamishi | 7 February 2024 sunset to 8 February 2024 sunset; Wednesday sunset until Thursday sunset

[5.Tag: 7.Februar 2024 Sonnenuntergang bis 8.Februar 2024 Sonnenuntergang; Mittwoch Sonnenuntergang bis Donnerstag Sonnenuntergang]

27

6th Day | Yom Shishi | 8 February 2024 sunset to 9 February 2024 sunset; Thursday sunset until Friday sunset: Old Light

[6.Tag: 8.Februar 2024 Sonnenuntergang bis 9.Februar 2024 Sonnenuntergang; Donnerstag Sonnenuntergang bis Freitag Sonnenuntergang: Altlicht]

28

7th Day | Shabbat | 9 February 2024 sunset to 10 February 2024 sunset; Friday sunset until Saturday sunset: New Moon

[7.Tag: 9.Februar 2024 Sonnenuntergang bis 10.Februar 2024 Sonnenuntergang; Freitag Sonnenuntergang bis Samstag Sonnenuntergang: Neumond]

29

1st Day | Yom Rishon | Rosh Chodesh: First Crescent Moon, sunset 11 Feb: New Light | 10 February 2024 sunset to 11 February 2024 sunset; Saturday sunset until Sunday sunset

[1.Tag: Rosh Chodesh: Erste sichtbare Mondsichel, Sonnenuntergang 11.Feb: Neulichtt | 10.Februar 2024 Sonnenuntergang bis 11.Februar 2024 Sonnenuntergang; Samstag Sonnenuntergang bis Sonntag Sonnenuntergang]

30 

Month 12 | Monat 12

2nd Day | Yom Sheni | Rosh Chodesh: New Light | 11 February 2024 sunset to 12 February 2024 sunset; Sunday sunset until Monday sunset

[2.Tag: Rosh Chodesh: Neulicht | 11.Februar 2024 Sonnenuntergang bis 12.Februar 2024 Sonnenuntergang; Sonntag Sonnenuntergang bis Montag Sonnenuntergang]

1 

3rd Day | Yom Shlishi | 12 February 2024 sunset to 13 February 2024 sunset; Monday sunset until Tuesday sunset

[3.Tag: 12.Februar 2024 Sonnenuntergang bis 13.Februar 2024 Sonnenuntergang; Montag Sonnenuntergang bis Dienstag Sonnenuntergang]

2

4th Day | Yom Revi í | 13 February 2024 sunset to 14 February 2024 sunset; Tuesday sunset until Wednesday sunset

[4.Tag: 13.Februar 2024 Sonnenuntergang bis 14.Februar 2024 Sonnenuntergang; Dienstag Sonnenuntergang bis Mittwoch Sonnenuntergang]

3

5th Day | Yom Hamishi | 14 February 2024 sunset to 15 February 2024 sunset; Wednesday sunset until Thursday sunset

[5.Tag: 14.Februar 2024 Sonnenuntergang bis 15.Februar 2024 Sonnenuntergang; Mittwoch Sonnenuntergang bis Donnerstag Sonnenuntergang]

6th Day | Yom Shishi | 15 February 2024 sunset to 16 February 2024 sunset; Thursday sunset until Friday sunset

[6.Tag: 15.Februar 2024 Sonnenuntergang bis 16.Februar 2024 Sonnenuntergang; Donnerstag Sonnenuntergang bis Freitag Sonnenuntergang]

5

7th Day | Shabbat | 16 February 2024 sunset to 17 February 2024 sunset; Friday sunset until Saturday sunset

[7.Tag: 16.Februar 2024 Sonnenuntergang bis 17.Februar 2024 Sonnenuntergang; Freitag Sonnenuntergang bis Samstag Sonnenuntergang]

6

1st Day | Yom Rishon | 17 February 2024 sunset to 18 February 2024 sunset; Saturday sunset until Sunday sunset

[1.Tag: 17.Februar 2024 Sonnenuntergang bis 18.Februar 2024 Sonnenuntergang; Samstag Sonnenuntergang bis Sonntag Sonnenuntergang]

7

2nd Day | Yom Sheni | 18 February 2024 sunset to 19 February 2024 sunset; Sunday sunset until Monday sunset

[2.Tag: 18.Februar 2024 Sonnenuntergang bis 19.Februar 2024 Sonnenuntergang; Sonntag Sonnenuntergang bis Montag Sonnenuntergang]

8

3rd Day | Yom Shlishi | 19 February 2024 sunset to 20 February 2024 sunset; Monday sunset until Tuesday sunset

[3.Tag: 19.Februar 2024 Sonnenuntergang bis 20.Februar 2024 Sonnenuntergang; Montag Sonnenuntergang bis Dienstag Sonnenuntergang]

9

4th Day | Yom Revi´i | 20 February 2024 sunset to 21 February 2024 sunset; Tuesday sunset until Wednesday sunset

[4.Tag: 20.Februar 2024 Sonnenuntergang bis 21.Februar 2024 Sonnenuntergang; Dienstag Sonnenuntergang bis Mittwoch Sonnenuntergang]

10

5th Day | Yom Hamishi | Ta´anit Esther: Commemorative Fast of Esther (Esther 9:1-14; Megillat 2A, Usually Observed Day 13 of Month 12) | 21 February 2024 sunset to 22 February 2024 sunset; Wednesday sunset until Thursday sunset

[5.Tag: Ta´anit Ester: Ester-Gedenkfasten (Ester 9,1-14; Megillat 2A, Gewöhnlich am 13.Tag des 12.Monats) | 21.Februar 2024 Sonnenuntergang bis 22.Februar 2024 Sonnenuntergang; Mittwoch Sonnenuntergang bis Donnerstag Sonnenuntergang]

11

6th Day | Yom Shishi | 22 February 2024 sunset to 23 February 2024 sunset; Thursday sunset until Friday sunset

[6.Tag: 22.Februar 2024 Sonnenuntergang bis 23.Februar 2024 Sonnenuntergang; Donnerstag Sonnenuntergang bis Freitag Sonnenuntergang]

12

7th Day | Shabbat | 23 February 2024 sunset to 24 February 2024 sunset; Friday sunset until Saturday sunset

[7.Tag: 23.Februar 2024 Sonnenuntergang bis 24.Februar 2024 Sonnenuntergang; Freitag Sonnenuntergang bis Samstag Sonnenuntergang]

13

1st Day | Yom Rishon | Purim Lots (Esther 9:15-17) | 24 February 2024 sunset to 25 February 2024 sunset; Saturday sunset until Sunday sunset: Full Moon

[1.Tag: Purim Lose (Ester 9,15-17) | 24.Februar 2024 Sonnenuntergang bis 25.Februar 2024 Sonnenuntergang; Samstag Sonnenuntergang bis Sonntag Sonnenuntergang: Vollmond]

14

2nd Day | Yom Sheni | Shushan Purim Lots (Esther 9:18-32) | 25 February 2024 sunset to <u>26 February</u> 2024 sunset; Sunday sunset until <u>Monday</u> sunset

[2.Tag: Shushan Purim Lots/Purim Lose der Städte (Ester 9,18-32) | 25.Februar 2024 Sonnenuntergang bis <u>26.Februar</u> 2024 Sonnenuntergang; Sonntag Sonnenuntergang bis <u>Montag</u> Sonnenuntergang]

15

3rd Day | Yom Shlishi | 26 February 2024 sunset to 27 February 2024 sunset; Monday sunset until Tuesday sunset

[3.Tag: 26.Februar 2024 Sonnenuntergang bis 27.Februar 2024 Sonnenuntergang; Montag Sonnenuntergang bis Dienstag Sonnenuntergang]

16

4th Day | Yom Revi´i | 27 February 2024 sunset to 28 February 2024 sunset; Tuesday sunset until Wednesday sunset

[4.Tag: 27.Februar 2024 Sonnenuntergang bis 28.Februar 2024 Sonnenuntergang; Dienstag Sonnenuntergang bis Mittwoch Sonnenuntergang]

17

5th Day | Yom Hamishi | 28 February 2024 sunset to 29 February 2024 sunset; Wednesday sunset until Thursday sunset

[5.Tag: 28.Februar 2024 Sonnenuntergang bis 29.Februar 2024 Sonnenuntergang; Mittwoch Sonnenuntergang bis Donnerstag Sonnenuntergang]

18

6th Day | Yom Shishi | 29 February 2024 sunset to 1 March 2024 sunset; Thursday sunset until Friday sunset

[6.Tag: 29.Februar 2024 Sonnenuntergang bis 1.März 2024 Sonnenuntergang; Donnerstag Sonnenuntergang bis Freitag Sonnenuntergang]

19

7th Day | Shabbat | 1 March 2024 sunset to 2 March 2024 sunset; Friday sunset until Saturday sunset

[7.Tag: 1.März 2024 Sonnenuntergang bis 2.März 2024 Sonnenuntergang; Freitag Sonnenuntergang bis Samstag Sonnenuntergang]

20

1st Day | Yom Rishon | 2 March 2024 sunset to 3 March 2024 sunset; Saturday sunset until Sunday sunset

[1.Tag: 2.März 2024 Sonnenuntergang bis 3.März 2024 Sonnenuntergang; Samstag Sonnenuntergang bis Sonntag Sonnenuntergang]

21

2nd Day | Yom Sheni | 3 March 2024 sunset to 4 March 2024 sunset; Sunday sunset until Monday sunset

[2.Tag: 3.März 2024 Sonnenuntergang bis 4.März 2024 Sonnenuntergang; Sonntag Sonnenuntergang bis Montag Sonnenuntergang]

22

3rd Day | Yom Shlishi | 4 March 2024 sunset to 5 March 2024 sunset; Monday sunset until Tuesday sunset

[3.Tag: 4.März 2024 Sonnenuntergang bis 5.März 2024 Sonnenuntergang; Montag Sonnenuntergang bis Dienstag Sonnenuntergang]

23

4th Day | Yom Revi´i | 5 March 2024 sunset to 6 March 2024 sunset; Tuesday sunset until Wednesday sunset

[4.Tag: 5.März 2024 Sonnenuntergang bis 6.März 2024 Sonnenuntergang; Dienstag Sonnenuntergang bis Mittwoch Sonnenuntergang]

24

5th Day | Yom Hamishi | 6 March 2024 sunset to 7 March 2024 sunset; Wednesday sunset until Thursday sunset

[5.Tag: 6.März 2024 Sonnenuntergang bis 7.März 2024 Sonnenuntergang; Mittwoch Sonnenuntergang bis Donnerstag Sonnenuntergang]

25

6th Day | Yom Shishi | 7 March 2024 sunset to 8 March 2024 sunset; Thursday sunset until Friday sunset

[6.Tag: 7.März 2024 Sonnenuntergang bis 8.März 2024 Sonnenuntergang; Donnerstag Sonnenuntergang bis Freitag Sonnenuntergang]

26

7th Day | Shabbat/Shabbat Ha-Kodesh | 8 March 2024 sunset to 9 March 2024 sunset; Friday sunset until Saturday sunset: Old Light

[7.Tag: 8.März 2024 Sonnenuntergang bis 9.März 2024 Sonnenuntergang; Freitag Sonnenuntergang bis Samstag Sonnenuntergang: Altlicht]

27

1st Day | Yom Rishon | 9 March 2024 sunset to 10 March 2024 sunset; Saturday sunset until Sunday sunset: New Moon

[1.Tag: 9.März 2024 Sonnenuntergang bis 10.März 2024 Sonnenuntergang; Samstag Sonnenuntergang bis Sonntag Sonnenuntergang: Neumond]

28 

2nd Day | Yom Sheni | Rosh Chodesh: First Crescent Moon sunset, 11 March: New Light | 10 March 2024 sunset to 11 March 2024 sunset; Sunday sunset until Monday sunset

[2.Tag: Rosh Chodesh: Erste sichtbare Mondsichel Sonnenuntergang 11.März: Neulicht | 10.März 2024 Sonnenuntergang bis 11.März 2024 Sonnenuntergang; Sonntag Sonnenuntergang bis Montag Sonnenuntergang]

29

Notes | Notizen

What do I want to change about myself, about my life for tomorrow's Spiritual Official New Year („Abib“)? What are my resolutions? How can I improve my faith life?

Was möchte ich für das geistige offizielle morgen beginnende Neujahr („Abib") an mir, an meinem Leben ändern? Was sind meine Vorsätze? Wie kann ich mein Glaubensleben verbessern?

Month 1 / *Monat 1*

Abib

Barley ripening (ripening of ears)

Gerstenreife (Ährenreife)

Notes | Notizen

3rd Day | Yom Shlishi | 11 March 2024 sunset to 12 March 2024 sunset; Monday sunset until Tuesday sunset

[3. Tag; 11.März 2024 Sonnenuntergang bis 12. März 2024 Sonnenuntergang; Montag Sonnenuntergang bis Dienstag Sonnenuntergang]

1 

Rosh Chodeshim

Spiritual New Year (Exodus 12:1-2) by Elohim

Geistiges Neujahr (2.Mose 12,1-2) von Elohim

4th Day | Yom Revi´i | 12 March 2024 sunset to 13 March 2024 sunset; Tuesday sunset until Wednesday sunset

[4.Tag: 12.März 2024 Sonnenuntergang bis 13.März 2024 Sonnenuntergang; Dienstag Sonnenuntergang bis Mittwoch Sonnenuntergang]

2

5th Day | Yom Hamishi | 13 March 2024 sunset to 14 March 2024 sunset; Wednesday sunset until Thursday sunset
[5.Tag: 13.März 2024 Sonnenuntergang bis 14.März 2024 Sonnenuntergang; Mittwoch Sonnenuntergang bis Donnerstag Sonnenuntergang]

3

6th Day | Yom Shishi | 14 March 2024 sunset to 15 March 2024 sunset; Thursday sunset until Friday sunset

[6.Tag: 14.März 2024 Sonnenuntergang bis 15.März 2024 Sonnenuntergang; Donnerstag Sonnenuntergang bis Freitag Sonnenuntergang]

4

7th Day | Shabbat | 15 March 2024 sunset to 16 March 2024 sunset; Friday sunset until Saturday sunset

[7.Tag: 15.März 2024 Sonnenuntergang bis 16.März 2024 Sonnenuntergang; Freitag Sonnenuntergang bis Samstag Sonnenuntergang]

5

1st Day | Yom Rishon | 16 March 2024 sunset to 17 March 2024 sunset; Saturday sunset until Sunday sunset

[1.Tag: 16.März 2024 Sonnenuntergang bis 17.März 2024 Sonnenuntergang; Samstag Sonnenuntergang bis Sonntag Sonnenuntergang]

6

2nd Day | Yom Sheni | 17 March 2024 sunset to 18 March 2024 sunset; Sunday sunset until Monday sunset

[2.Tag: 17.März 2024 Sonnenuntergang bis 18.März 2024 Sonnenuntergang; Sonntag Sonnenuntergang bis Montag Sonnenuntergang]

7

3rd Day | Yom Shlishi | 18 March 2024 sunset to 19 March 2024 sunset; Monday sunset until Tuesday sunset

[3.Tag: 18.März 2024 Sonnenuntergang bis 19.März 2024 Sonnenuntergang; Montag Sonnenuntergang bis Dienstag Sonnenuntergang]

8

4th Day | Yom Revi ´i | Spring Equinox: 20 March, 03h:06m UT | 19 March 2024 sunset to 20 March 2024 sunset; Tuesday sunset until Wednesday sunset

[4.Tag: Frühlingsäquinoktium/-Tag-und-Nacht-Gleiche: 20.März, 03Std:06min UT (Weltzeit) | 19.März 2024 Sonnenuntergang bis 20.März 2024 Sonnenuntergang; Dienstag Sonnenuntergang bis Mittwoch Sonnenuntergang]

9

5th Day | Yom Hamishi | Fast of Nadab and Abihu (Leviticus 10:1-2) | 20 March 2024 sunset to 21 March 2024 sunset; Wednesday sunset until Thursday sunset

[5.Tag: Nadab und Abihu Fasten (Levitikus/3.Mose 10,1-2); 20.März 2024 Sonnenuntergang bis 21.März 2024 Sonnenuntergang; Mittwoch Sonnenuntergang bis Donnerstag Sonnenuntergang]

10

6th Day | Yom Shishi | Fast of Ezra (Ezra´s Aliyah to Jerusalem, Ezra 8:21-31) | 21 March 2024 sunset to 22 March 2024 sunset; Thursday sunset until Friday sunset

[6.Tag: Ezra-Fasten (Ezras Aliyah nach Jerusalem, Ezra 8,21-31) | 21.März 2024 Sonnenuntergang bis 22.März 2024 Sonnenuntergang; Donnerstag Sonnenuntergang bis Freitag Sonnenuntergang]

11

7th Day | Shabbat | 22 March 2024 sunset to 23 March 2024 sunset; Friday sunset until Saturday sunset

[7.Tag: 22.März 2024 Sonnenuntergang bis 23.März 2024 Sonnenuntergang; Freitag Sonnenuntergang bis Samstag Sonnenuntergang]

12

1st Day | Yom Rishon | Fast of the Firstborn (Soferim 21:3) | Original Fast of Esther 4:16 | Prepare for the Seder Meal | 23 March 2024 sunset to 24 March 2024 sunset; Saturday sunset until Sunday sunset

[1.Tag: Fasten der Erstgeborenen (Soferim 21:3) | Original Ester-Fasten (Ester 4,16) | Vorbereitung Seder-Mahlzeit | 23.März 2024 Sonnenuntergang bis 24.März 2024 Sonnenuntergang; Samstag Sonnenuntergang bis Sonntag Sonnenuntergang]

13

2nd Day |Yom Sheni | Pesach: Passover , Day of Redemption, The Lamb of Elohim Slain | Original Fast of Esther 4:16 | Penumbral Lunar Eclipse | Full Moon | 24 March 2024 sunset to 25 March 2024 sunset; Sunday sunset until Monday sunset

[2.Tag: Pesach (Tag der Erlösung: Das Lamm Elohims wird geschlachtet); Original Ester-Fasten (Ester 4,16); Totale Halbschatten-Mondfinsternis |Vollmond | 24.März 2024 Sonnenuntergang bis 25.März 2024 Sonnenuntergang; Sonntag Sonnenuntergang bis Montag Sonnenuntergang]

14  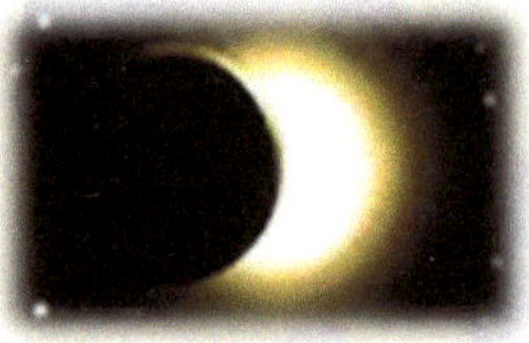

3rd Day | Yom Shlishi | Hag HaMatzah: Unleavened Bread 1st Day: Israel Leaves Rameses; Passover Day 1 | Original Fast of Esther 4:16 | 25 March 2024 sunset to 26 March 2024 sunset; Monday sunset until Tuesday sunset

[3.Tag: 1.Tag Fest der Ungesäuerten Brote (Hag HaMatza): Israel verlässt Rameses; Pesach 1.Tag | Original Ester-Fasten (Ester 4,16); 25.März 2024 Sonnenuntergang bis 26.März 2024 Sonnenuntergang; Montag Sonnenuntergang bis Dienstag Sonnenuntergang]

15

4th Day | Hag HaMatzah: Unleavened Bread, Resheet – First Fruits: The Omer Offering (Omer Day 1); Passover Day 2 | 26 March 2024 sunset to 27 March 2024 sunset; Tuesday sunset until Wednesday sunset

[4.Tag: Hag HaMatza: Ungesäuerte Brote, Resheet – Erstlingsfrüchte: Das Omer-Opfer (Omer Tag 1); Pesach Tag 2 | 26.März 2024 Sonnenuntergang bis 27.März 2024 Sonnenuntergang; Dienstag Sonnenuntergang bis Mittwoch Sonnenuntergang]

16

5th Day | Yom Hamishi | Hag HaMatzah: Unleavened Bread: The Risen Lamb of Elohim; Passover Day 3; Omer Day 2 | 27 March 2024 sunset to 28 March 2024 sunset; Wednesday sunset until Thursday sunset

[5.Tag: Hag HaMatza: Ungesäuerte Brote: Das Auferstandene Lamm Elohims; Pesach Tag 3; Omer Tag 2 | 27.März 2024 Sonnenuntergang bis 28.März 2024 Sonnenuntergang; Mittwoch Sonnenuntergang bis Donnerstag Sonnenuntergang]

17

6th Day | Yom Shishi | Hag HaMatzah: Unleavened Bread, Passover Day 4, Omer Day 3 | 28 March 2024 sunset to 29 March 2024 sunset; Thursday sunset until Friday sunset

[6.Tag: Hag HaMatza: Ungesäuerte Brote, Pesach Tag 4, Omer Tag 3 | 28.März 2024 Sonnenuntergang bis 29.März 2024 Sonnenuntergang; Donnerstag Sonnenuntergang bis Freitag Sonnenuntergang]

18

7th Day | Shabbat | Hag HaMatzah: Unleavened Bread: Passover Day Shabbat; 1st Shabbat of Omer; Omer Day 4 | 29 March 2024 sunset to 30 March 2024 sunset; Friday sunset until Saturday sunset

[7.Tag: Hag HaMatza: Ungesäuerte Brote: Pesach Shabbat Tag; 1.Shabbat des Omer; Omer Tag 4 | 29.März 2024 Sonnenuntergang bis 30.März 2024 Sonnenuntergang; Freitag Sonnenuntergang bis Samstag Sonnenuntergang]

19

1st Day | Yom Rishon | Hag HaMatzah: Unleavened Bread: Passover Day 6: Omer Day 5 | 30 March 2024 sunset to 31 March 2024 sunset; Saturday sunset until Sunday sunset
[1.Tag: Hag HaMatza: Ungesäuerte Brote: Pesach Tag 6; Omer Tag 5 | 30.März 2024 Sonnenuntergang bis 31.März 2024 Sonnenuntergang; Samstag Sonnenuntergang bis Sonntag Sonnenuntergang]

20

2nd Day | Yom Sheni | Hag HaMatzah: Unleavened Bread 7th Day: Israel Crosses the Red Sea, Passover Day 7, A Solemn Assembly (Deuteronomy 16:8), Omer Day 6 | 31 March 2024 sunset to 1 April 2024 sunset; Sunday sunset until Monday sunset

[2.Tag: Hag HaMatza: Ungesäuerte Brote 7.Tag: Israel überquert das Rote Meer, Pesach Tag 7, Eine feierliche Versammlung (5.Mose 16,8), Omer Tag 6 | 31.März 2024 Sonnenuntergang bis 1.April 2024 Sonnenuntergang; Sonntag Sonnenuntergang bis Montag Sonnenuntergang]

21

3rd Day | Yom Shlishi | Omer Day 7 | 1 April 2024 sunset to 2 April 2024 sunset; Monday sunset until Tuesday sunset

[3.Tag: Omer Tag 7, 1.April 2024 Sonnenuntergang bis 2.April 2024 Sonnenuntergang; Montag Sonnenuntergang bis Dienstag Sonnenuntergang]

22

4th Day | Yom Revi ´i | Omer Day 8 | 2 April 2024 sunset to 3 April 2024 sunset; Tuesday sunset until Wednesday sunset

[4.Tag: Omer Tag 8 | 2.April 2024 Sonnenuntergang bis 3.April 2024 Sonnenuntergang; Dienstag Sonnenuntergang bis Mittwoch Sonnenuntergang]

23

5th Day | Yom Hamishi | Omer Day 9 | 3 April 2024 sunset to 4 April 2024 sunset; Wednesday sunset until Thursday sunset

[5.Tag: Omer Tag 9 | 3.April 2024 Sonnenuntergang bis 4.April 2024 Sonnenuntergang; Mittwoch Sonnenuntergang bis Donnerstag Sonnenuntergang]

24

6th Day | Yom Shishi | Omer Day 10 | 4 April 2024 sunset to 5 April 2024 sunset; Thursday sunset until Friday sunset

[6.Tag: Omer Tag 10 | 4.April 2024 Sonnenuntergang bis 5.April 2024 Sonnenuntergang; Donnerstag Sonnenuntergang bis Freitag Sonnenuntergang]

25

7th Day | Shabbat | Omer Day 11, 2nd Shabbat | 5 April 2024 sunset to 6 April 2024 sunset; Friday sunset until Saturday sunset

[7.Tag: Omer Tag 11, 2.Shabbat | 5.April 2024 Sonnenuntergang bis 6.April 2024 Sonnenuntergang; Freitag Sonnenuntergang bis Samstag Sonnenuntergang]

26

1st Day | Yom Rishon | Omer Day 12 | 6 April 2024 sunset to 7 April 2024 sunset; Saturday sunset until Sunday sunset: Old Light

[1.Tag: Omer Tag 12 | 6.April 2024 Sonnenuntergang bis 7.April 2024 Sonnenuntergang; Samstag Sonnenuntergang bis Sonntag Sonnenuntergang: Altlicht]

27

2nd Day | Omer Day 13 | Yom HaShoah: Israel State Holiday Holocaust Remembrance Day, Usually Observed Day 27 of Month 1 | 7 April 2024 sunset to 8 April 2024 sunset; Sunday sunset until Monday sunset: New Moon

[2.Tag: Omer Tag 13 | Yom HaShoah: Israelischer nationaler Gedenktag für die Opfer der Schoah, gewöhnlich am 27.Tag des 1.Monats | 7.April 2024 Sonnenuntergang bis 8.April 2024 Sonnenuntergang; Sonntag Sonnenuntergang bis Montag Sonnenuntergang: Neumond]

28 

3rd Day | Yom Shlishi | Rosh Chodesh: First Crescent Moon sunset, 9 April | Total Solar Eclipse | New Light | Omer Day 14 | 8 April 2024 sunset to 9 April 2024 sunset; Monday sunset until Tuesday sunset

[3.Tag: Rosh Chodesh: Erste sichtbare Mondsichel Sonnenuntergang am 9.April | Totale Sonnenfinsternis | Neulicht | Omer Tag 14 | 8.April 2024 Sonnenuntergang bis 9.April 2024 Sonnenuntergang; Montag Sonnenuntergang bis Dienstag Sonnenuntergang]

29  

Notes | Notizen

Month 2 | *Monat 2*

Notes | Notizen

4th Day | Rosh Chodesh: New Light | Omer Day 15 | 9 April 2024 sunset to 10 April 2024 sunset; Tuesday sunset until Wednesday sunset

[4.Tag: Rosh Chodesh: Neulicht | Omer Tag 15 | 9.April 2024 Sonnenuntergang bis 10.April 2024 Sonnenuntergang, Dienstag Sonnenuntergang bis Mittwoch Sonnenuntergang]

1

5th Day | Yom Hamishi | Omer Day 16 | 10 April 2024 sunset to 11 April 2024 sunset; Wednesday sunset until Thursday sunset

[5.Tag: Omer Tag 16 | 10.April 2024 Sonnenuntergang bis 11.April 2024 Sonnenuntergang; Mittwoch Sonnenuntergang bis Donnerstag Sonnenuntergang]

2

6th Day | Yom Shishi | Omer Day 17 | 11 April 2024 sunset to 12 April 2024 sunset; Thursday sunset until Friday sunset

[6.Tag: Omer Tag 17 | 11.April 2024 Sonnenuntergang bis 12.April 2024 Sonnenuntergang; Donnerstag Sonnenuntergang bis Freitag Sonnenuntergang]

3

7th Day | Shabbat | Omer Day 18 | 3rd Shabbat of Omer | 12 April 2024 sunset to 13 April 2024 sunset; Friday sunset until Saturday sunset

[7.Tag: Omer Tag 18; 3.Shabbat des Omer | 12.April 2024 Sonnenuntergang bis 13.April 2024 Sonnenuntergang; Freitag Sonnenuntergang bis Samstag Sonnenuntergang]

4

1st Day | Yom Rishon | Omer Day 19 | 13 April 2024 sunset to 14 April 2024 sunset; Saturday sunset until Sunday sunset

[1.Tag: Omer Tag 19 | 13.April 2024 Sonnenuntergang bis 14.April 2024 Sonnenuntergang; Samstag Sonnenuntergang bis Sonntag Sonnenuntergang]

5

2nd Day | Yom Sheni | Omer Day 20 | 14 April 2024 sunset to 15 April 2024 sunset; Sunday sunset until Monday sunset

[2.Tag: Omer Tag 20 | 14.April 2024 Sonnenuntergang bis 15.April 2024 Sonnenuntergang; Sonntag Sonnenuntergang bis Montag Sonnenuntergang]

6

3rd Day | Yom Shlishi | Omer Day 21 | 15 April 2024 sunset to 16 April 2024 sunset; Monday sunset until Tuesday sunset

[3.Tag: Omer Tag 21 | 15.April 2024 Sonnenuntergang bis 16.April 2024 Sonnenuntergang; Montag Sonnenuntergang bis Dienstag Sonnenuntergang]

7

4th Day | Yom Revi ´i | Omer Day 22 | 16 April 2024 sunset to 17 April 2024 sunset; Tuesday sunset until Wednesday sunset

[4.Tag: Omer Tag 22 | 16.April 2024 Sonnenuntergang bis 17.April 2024 Sonnenuntergang; Dienstag Sonnenuntergang bis Mittwoch Sonnenuntergang]

8

5th Day | Yom Hamishi | Omer Day 23 | 17 April 2024 sunset to 18 April 2024 sunset; Wednesday sunset until Thursday sunset

[5.Tag: Omer Tag 23 | 17.April 2024 Sonnenuntergang bis 18.April 2024 Sonnenuntergang; Mittwoch Sonnenuntergang bis Donnerstag Sonnenuntergang]

9

6th Day | Yom Shishi | Omer Day 24 | 18 April 2024 sunset to 19 April 2024 sunset; Thursday sunset until Friday sunset

[6.Tag: Omer Tag 24 | 18.April 2024 Sonnenuntergang bis 19.April 2024 Sonnenuntergang; Donnerstag Sonnenuntergang bis Freitag Sonnenuntergang]

10

7th Day | Shabbat | Omer Day 25; 4th Shabbat of Omer | 19 April 2024 sunset to 20 April 2024 sunset; Friday sunset until Saturday sunset

[7.Tag: Omer Tag 25; 4.Shabbat des Omer | 19.April 2024 Sonnenuntergang bis 20.April 2024 Sonnenuntergang; Freitag Sonnenuntergang bis Samstag Sonnenuntergang]

11

1st Day | Yom Rishon | Omer Day 26 | 20 April 2024 sunset to 21 April 2024 sunset; Saturday sunset until Sunday sunset

[1.Tag: Omer Tag 26 | 20.April 2024 Sonnenuntergang bis 21.April 2024 Sonnenuntergang; Samstag Sonnenuntergang bis Sonntag Sonnenuntergang]

12

2nd Day | Yom Sheni | Omer Day 27 | 21 April 2024 sunset to 22 April 2024 sunset; Sunday sunset until Monday sunset

[2.Tag: Omer Tag 27 | 21.April 2024 Sonnenuntergang bis 22.April 2024 Sonnenuntergang; Sonntag Sonnenuntergang bis Montag Sonnenuntergang]

13

3rd Day | Yom Shlishi | Omer Day 28 | Pesach Sheni; Second Passover (Numbers 9:6-14 | 2 Chronicles 29:1-27) | 22 April 2024 sunset to 23 April 2024 sunset; Monday sunset until Tuesday sunset; Full Moon

[3.Tag: Omer Tag 28 | Pesach Sheni; Zweites Pesach (Numeri/4.Mose 9,6-14 | 2.Chronik 29,1-27) | 22.April 2024 Sonnenuntergang bis 23.April 2024 Sonnenuntergang; Montag Sonnenuntergang bis Dienstag Sonnenuntergang; Vollmond]

14 

4th Day | Yom Revi ´i | Omer Day 29 | 23 April 2024 sunset to 24 April 2024 sunset; Tuesday sunset until Wednesday sunset

[4.Tag: Omer Tag 29 | 23.April 2024 Sonnenuntergang bis 24.April 2024 Sonnenuntergang; Dienstag Sonnenuntergang bis Mittwoch Sonnenuntergang]

15

5th Day | Yom Hamishi | Omer Day 30 | 24 April 2024 sunset to 25 April 2024 sunset; Wednesday sunset until Thursday sunset

[5.Tag: Omer Tag 30 | 24.April 2024 Sonnenuntergang bis 25.April 2024 Sonnenuntergang; Mittwoch Sonnenuntergang bis Donnerstag Sonnenuntergang]

16

6th Day | Yom Shishi | Omer Day 31 | 25 April 2024 sunset to 26 April 2024 sunset; Thursday sunset until Friday sunset

[6.Tag: Omer Tag 31 | 25.April 2024 Sonnenuntergang bis 26.April 2024 Sonnenuntergang; Donnerstag Sonnenuntergang bis Freitag Sonnenuntergang]

17

7th Day | Shabbat | Omer Day 32; 5th Shabbat of Omer | 26 April 2024 sunset to 27 April 2024 sunset; Friday sunset until Saturday sunset

[7.Tag: Omer Tag 32; 5.Shabbat des Omer | 26.April 2024 Sonnenuntergang bis 27.April 2024 Sonnenuntergang; Freitag Sonnenuntergang bis Samstag Sonnenuntergang]

18

1st Day | Yom Rishon | Omer Day 33 | 27 April 2024 sunset to 28 April 2024 sunset; Saturday sunset until Sunday sunset

[1.Tag: Omer Tag 33 | 27.April 2024 Sonnenuntergang bis 28.April 2024 Sonnenuntergang; Samstag Sonnenuntergang bis Sonntag Sonnenuntergang]

19

2nd Day | Yom Sheni | Omer Day 34 | 28 April 2024 sunset to 29 April 2024 sunset; Sunday sunset until Monday sunset

[2.Tag: Omer Tag 34 | 28.April 2024 Sonnenuntergang bis 29.April 2024 Sonnenuntergang; Sonntag Sonnenuntergang bis Montag Sonnenuntergang]

20

3rd Day | Yom Shlishi | Omer Day 35 | 29 April 2024 sunset to 30 April 2024 sunset; Monday sunset until Tuesday sunset

[3.Tag: Omer Tag 35 | 29.April 2024 Sonnenuntergang bis 30.April 2024 Sonnenuntergang; Montag Sonnenuntergang bis Dienstag Sonnenuntergang]

21

4th Day | Yom Revi ´i | Omer Day 36 | 30 April 2024 sunset to 1 May 2024 sunset; Tuesday sunset until Wednesday sunset

[4.Tag: Omer Tag 36 | 30.April 2024 Sonnenuntergang bis 1.Mai 2024 Sonnenuntergang; Dienstag Sonnenuntergang bis Mittwoch Sonnenuntergang]

22

5th Day | Yom Hamishi | Omer Day 37 | 1 May 2024 sunset to 2 May 2024 sunset; Wednesday sunset until Thursday sunset

[5.Tag: Omer Tag 37 | 1.Mai 2024 Sonnenuntergang bis 2.Mai 2024 Sonnenuntergang; Mittwoch Sonnenuntergang bis Donnerstag Sonnenuntergang]

23

6th Day | Yom Shishi | Omer Day 38 | 2 May 2024 sunset to 3 May 2024 sunset; Thursday sunset until Friday sunset

[6.Tag: Omer Tag 38 | 2.Mai 2024 Sonnenuntergang bis 3.Mai 2024 Sonnenuntergang; Donnerstag Sonnenuntergang bis Freitag Sonnenuntergang]

24

7th Day | Shabbat | Omer Day 39; 6th Shabbat of Omer | 3 May 2024 sunset to 4 May 2024 sunset; Friday sunset until Saturday sunset

[7.Tag: Omer Tag 39; 6.Shabbat im Omer | 3.Mai 2024 Sonnenuntergang bis 4.Mai 2024 Sonnenuntergang; Freitag Sonnenuntergang bis Samstag Sonnenuntergang]

25

1st Day | Yom Rishon | Omer Day 40 | 4 May 2024 sunset to 5 May 2024 sunset; Saturday sunset until Sunday sunset

[1.Tag: Omer Tag 40 | 4.Mai 2024 Sonnenuntergang bis 5.Mai 2024 Sonnenuntergang; Samstag Sonnenuntergang bis Sonntag Sonnenuntergang]

26

2nd Day | Yom Sheni | Omer Day 41 | Aliyah of Messiah (Mark 16:19-20 | Luke 24:50-53 | Acts 1:1-14) | 5 May 2024 sunset to 6 May 2024 sunset; Sunday sunset until Monday sunset

[2.Tag: Omer Tag 41 | Aliyah des Messias (Markus 16,19-20 | Lukas 24,50-53 | Apostelgeschichte 1,1.14) | 5.Mai 2024 Sonnenuntergang bis 6.Mai 2024 Sonnenuntergang; Sonntag Sonnenuntergang bis Montag Sonnenuntergang]

27

3rd Day | Yom Shlishi | Omer Day 42 | Fast of Samuel: Death of Samuel (Megillat Ta´anit) | 6 May 2024 sunset to 7 May 2024 sunset; Monday sunset until Tuesday sunset: Old light

[3.Tag: Omer Tag 42 | Samuel-Fasten: Samuels Tod (Megillat Ta´anit) | 6.Mai 2024 Sonnenuntergang bis 7.Mai 2024 Sonnenuntergang; Montag Sonnenuntergang bis Dienstag Sonnenuntergang: Altlicht]

28

4th Day | Yom Revi´i | Omer Day 43 | 7 May 2024 sunset to 8 May 2024 sunset; Tuesday sunset until Wednesday sunset: New Moon

[4.Tag: Omer Tag 43 | 7.Mai 2024 Sonnenuntergang bis 8.Mai 2024 Sonnenuntergang; Dienstag Sonnenuntergang bis Mittwoch Sonnenuntergang: Neumond]

29

5th Day | Yom Hamishi | Omer Day 44 | Rosh Chodesh: First Crescent Moon sunset, 9 May: New Light | 8 May 2024 sunset to 9 May 2024 sunset; Wednesday sunset until Thursday sunset

[5.Tag: Omer Tag 44 | Rosh Chodesh: Erste sichtbare Mondsichel; Sonnenuntergang 9.Mai: Neulicht | 8.Mai 2024 Sonnenuntergang bis 9.Mai 2024 Sonnenuntergang; Mittwoch Sonnenuntergang bis Donnerstag Sonnenuntergang]

30 

Month 3| *Monat 3*

Notes | Notizen

6th Day | Yom Shishi | Omer Day 45 | Rosh Chodesh: New Light | 9 May 2024 sunset to 10 May 2024 sunset; Thursday sunset until Friday sunset

[6.Tag: Omer Tag 45 | Rosh Chodesh: Neulicht | 9.Mai 2024 Sonnenuntergang bis 10.Mai 2024 Sonnenuntergang; Donnerstag Sonnenuntergang bis Freitag Sonnenuntergang]

1

7th Day | Shabbat | Omer Day 46; 7th Shabbat of Omer | 10 May 2024 sunset to 11 May 2024 sunset; Friday sunset until Saturday sunset

[7.Tag: Omer Tag 46; 7.Shabbat im Omer | 10.Mai 2024 Sonnenuntergang bis 11.Mai 2024 Sonnenuntergang; Freitag Sonnenuntergang bis Samstag Sonnenuntergang]

2

1st Day | Yom Rishon | Omer Day 47 | 11 May 2024 sunset to 12 May 2024 sunset; Saturday sunset until Sunday sunset

[1.Tag: Omer Tag 47 | 11.Mai 2024 Sonnenuntergang bis 12.Mai 2024 Sonnenuntergang; Samstag Sonnenuntergang bis Sonntag Sonnenuntergang]

3

2nd Day | Yom Sheni | Omer Day 48 | Yom Ha-Atzma´ut: Israel State Holiday Independence Day | 12 May 2024 sunset to 13 May 2024 sunset; Sunday sunset until Monday sunset

[2.Tag: Omer Tag 48 | Yom Ha-Atzma´ut: Israelischer Nationalfeiertag der Unabhängigkeit | 12.Mai 2024 Sonnenuntergang bis 13.Mai 2024 Sonnenuntergang; Sonntag Sonnenuntergang bis Montag Sonnenuntergang]

4

3rd Day | Yom Shlishi | Omer Day 49 | Yom Ha-Atzma´ut: Israel State Holiday Independence Day, Usually Observed Day 4 of Month 3 | 13 May 2024 sunset to 14 May 2024 sunset; Monday sunset until Tuesday sunset

[3.Tag: Omer Day 49 | Yom Ha-Atzma´ut: Israelischer Nationalfeiertag der Unabhängigkeit, Gewöhnlich am 4.Tag des 3.Monats | 13.Mai 2024 Sonnenuntergang bis 14.Mai 2024 Sonnenuntergang; Montag Sonnenuntergang bis Dienstag Sonnenuntergang]

5

4th Day | Yom Revi´i | Omer Day 50: Last Day Counting the Omer | Shavout: Pentecost; Bikkurim – First Fruits, Spirit & Truth | 14 May 2024 sunset to 15 May 2024 sunset; Tuesday sunset until Wednesday sunset

[4.Tag: Omer Tag 50: Letzter Tag des Omerzählens | Shavout: Pfingsten; Bikkurim – Erstlingsfrüchte, Geist & Wahrheit | 14.Mai 2024 Sonnenuntergang bis 15.Mai 2024 Sonnenuntergang; Dienstag Sonnenuntergang bis Mittwoch Sonnenuntergang]

6

5th Day | Yom Hamishi | 15 May 2024 sunset to 16 May 2024 sunset; Wednesday sunset until Thursday sunset

[5.Tag: 15.Mai 2024 Sonnenuntergang bis 16.Mai 2024 Sonnenuntergang; Mittwoch Sonnenuntergang bis Donnerstag Sonnenuntergang]

7

6th Day | Yom Shishi | 16 May 2024 sunset to 17 May 2024 sunset; Thursday sunset until Friday sunset

[6.Tag: 16.Mai 2024 Sonnenuntergang bis 17.Mai 2024 Sonnenuntergang; Donnerstag Sonnenuntergang bis Freitag Sonnenuntergang]

8

7th Day | Shabbat | 17 May 2024 sunset to 18 May 2024 sunset; Friday sunset until Saturday sunset

[7.Tag: 17.Mai 2024 Sonnenuntergang bis 18.Mai 2024 Sonnenuntergang; Freitag Sonnenuntergang bis Samstag Sonnenuntergang]

9

1st Day | Yom Rishon | 18 May 2024 sunset to 19 May 2024 sunset; Saturday sunset until Sunday sunset

[1.Tag: 18.Mai 2024 Sonnenuntergang bis 19.Mai 2024 Sonnenuntergang; Samstag Sonnenuntergang bis Sonntag Sonnenuntergang]

10

2nd Day | Yom Sheni | 19 May 2024 sunset to 20 May 2024 sunset; Sunday sunset until Monday sunset

[2.Tag: 19.Mai 2024 Sonnenuntergang bis 20.Mai 2024 Sonnenuntergang; Sonntag Sonnenuntergang bis Montag Sonnenuntergang]

11

3rd Day | Yom Shlishi | 20 May 2024 sunset to <u>21 May</u> 2024 sunset; Monday sunset until <u>Tuesday</u> sunset

[3.Tag: 20.Mai 2024 Sonnenuntergang bis <u>21.Mai</u> 2024 Sonnenuntergang; Montag Sonnenuntergang bis <u>Dienstag</u> Sonnenuntergang]

12

4th Day | Yom Revi´i | 21 May 2024 sunset to 22 May 2024 sunset; Tuesday sunset until Wednesday sunset

[4.Tag: 21.Mai 2024 Sonnenuntergang bis 22.Mai 2024 Sonnenuntergang; Dienstag Sonnenuntergang bis Mittwoch Sonnenuntergang]

13

5th Day | Yom Hamishi | 22 May 2024 sunset to 23 May 2024 sunset; Wednesday sunset until Thursday sunset: Full Moon

[5.Tag: 22.Mai 2024 Sonnenuntergang bis 23.Mai 2024 Sonnenuntergang; Mittwoch Sonnenuntergang bis Donnerstag Sonnenuntergang: Vollmond]

14 

6th Day | Yom Shishi | 23 May 2024 sunset to 24 May 2024 sunset; Thursday sunset until Friday sunset

[6.Tag: 23.Mai 2024 Sonnenuntergang bis 24.Mai 2024 Sonnenuntergang; Donnerstag Sonnenuntergang bis Freitag Sonnenuntergang]

15

7th Day | Shabbat | 24 May 2024 sunset to 25 May 2024 sunset; Friday sunset until Saturday sunset

[7.Tag: 24.Mai 2024 Sonnenuntergang bis 25.Mai 2024 Sonnenuntergang; Freitag Sonnenuntergang bis Samstag Sonnenuntergang]

16

1st Day | Yom Rishon | 25 May 2024 sunset to 26 May 2024 sunset; Saturday sunset until Sunday sunset

[1.Tag: 25.Mai 2024 Sonnenuntergang bis 26.Mai 2024 Sonnenuntergang; Samstag Sonnenuntergang bis Sonntag Sonnenuntergang]

17

2nd Day | Yom Sheni | 26 May 2024 sunset to 27 May 2024 sunset; Sunday sunset until Monday sunset

[2.Tag: 26.Mai 2024 Sonnenuntergang bis 27.Mai 2024 Sonnenuntergang, Sonntag Sonnenuntergang bis Montag Sonnenuntergang]

18

3rd Day | Yom Shlishi | 27 May 2024 sunset to 28 May 2024 sunset; Monday sunset until Tuesday sunset

[3.Tag: 27.Mai 2024 Sonnenuntergang bis 28.Mai 2024 Sonnenuntergang; Montag Sonnenuntergang bis Dienstag Sonnenuntergang]

19

4th Day | Yom Revi ´i | 28 May 2024 sunset to 29 May 2024 sunset; Tuesday sunset until Wednesday sunset

[4.Tag: 28.Mai 2024 Sonnenuntergang bis 29.Mai 2024 Sonnenuntergang; Dienstag Sonnenuntergang bis Mittwoch Sonnenuntergang]

20

5th Day | Yom Hamishi | 29 May 2024 sunset to 30 May 2024 sunset; Wednesday sunset to Thursday sunset

[5.Tag: 29.Mai 2024 Sonnenuntergang bis 30.Mai 2024 Sonnenuntergang; Mittwoch Sonnenuntergang bis Donnerstag Sonnenuntergang]

21

6th Day | Yom Shishi | Fast of Jerobeam: For Jerobeam´s Sin of Leading the 10 Tribes Into Idolatry (1 Kings 12:32-33); Megillat Ta´anit, Usually Observed Day 23 of Month 3 | 30 May 2024 sunset to 31 May 2024 sunset; Thursday sunset until Friday sunset

[6.Tag: Jerobeam-Fasten: Für die Sünde Jerobeams, die 10 Stämme in Götzenanbetung geführt zu haben (1 Könige 12,32-33); Megillat Ta´anit, Gewöhnlich am 23.Tag des 3.Monats | 30.Mai 2024 Sonnenuntergang bis 31.Mai 2024 Sonnenuntergang; Donnerstag Sonnenuntergang bis Freitag Sonnenuntergang]

22

7th Day | Shabbat | 31 May 2024 sunset to 1 June 2024 sunset; Friday sunset until Saturday sunset

[7.Tag: 31.Mai 2024 Sonnenuntergang bis 1.Juni 2024 Sonnenuntergang; Freitag Sonnenuntergang bis Samstag Sonnenuntergang]

23

1st Day | Yom Rishon | 1 June 2024 sunset to 2 June 2024 sunset; Saturday sunset until Sunday sunset

[1.Tag: 1.Juni 2024 Sonnenuntergang bis 2.Juni 2024 Sonnenuntergang; Samstag Sonnenuntergang bis Sonntag Sonnenuntergang]

24

2nd Day | Yom Sheni | 2 June 2024 sunset to 3 June 2024 sunset; Sunday sunset until Monday sunset

[2.Tag: 2.Juni 2024 Sonnenuntergang bis 3.Juni 2024 Sonnenuntergang; Sonntag Sonnenuntergang bis Montag Sonnenuntergang]

25

3rd Day | Yom Shlishi | 3 June 2024 sunset to 4 June 2024 sunset; Monday sunset until Tuesday sunset

[3.Tag: 3.Juni 2024 Sonnenuntergang bis 4.Juni 2024 Sonnenuntergang; Montag Sonnenuntergang bis Dienstag Sonnenuntergang]

26

4th Day | Yom Revi ´i | 4 June 2024 sunset to 5 June 2024 sunset; Tuesday sunset until Wednesday sunset: Old Light

[4.Tag: 4.Juni 2024 Sonnenuntergang bis 5.Juni 2024 Sonnenuntergang; Dienstag Sonnenuntergang bis Mittwoch Sonnenuntergang: Altlicht]

27

5th Day | Yom Hamishi | Yom Yerushalayim: Israel State Holiday Jerusalem Day | 5 June 2024 sunset to 6 June 2024 sunset; Wednesday sunset until Thursday sunset: New Moon

[5.Tag: Yom Yerushalayim: Israelischer Nationalfeiertag Jerusalemtag | 5.Juni 2024 Sonnenuntergang bis 6.Juni 2024 Sonnenuntergang; Mittwoch Sonnenuntergang bis Donnerstag Sonnenuntergang: Neumond]

28 

6th Day | Yom Shishi | Rosh Chodesh: First Crescent Moon sunset, 7 June: New Light | 6 June 2024 sunset to 7 June 2024 sunset; Thursday sunset until Friday sunset

[6.Tag: Rosh Chodesh: Erste sichtbare Mondsichel, Sonnenuntergang 7.Juni: Neulicht | 6.Juni 2024 Sonnenuntergang bis 7.Juni 2024 Sonnenuntergang; Donnerstag Sonnenuntergang bis Freitag Sonnenuntergang]

29

Notes | Notizen

Month 4 | *Monat 4*

Notes | Notizen

7th Day | Shabbat | Rosh Chodesh: New Light | Shabbat Rosh Chodesh | 7 June 2024 sunset to 8 June 2024 sunset; Friday sunset until Saturday sunset

[7.Tag: Rosh Chodesh: Neulicht | Shabbat Rosh Chodesh | 7.Juni 2024 Sonnenuntergang bis 8.Juni 2024 Sonnenuntergang; Freitag Sonnenuntergang bis Samstag Sonnenuntergang]

1 

1st Day | Yom Rishon | 8 June 2024 sunset to 9 June 2024 sunset; Saturday sunset until Sunday sunset

[1.Tag: 8.Juni 2024 Sonnenuntergang bis 9.Juni 2024 Sonnenuntergang; Samstag Sonnenuntergang bis Sonntag Sonnenuntergang]

2

2nd Day | Yom Sheni | 9 June 2024 sunset to 10 June 2024 sunset; Sunday sunset until Monday sunset

[2.Tag: 9.Juni 2024 Sonnenuntergang bis 10.Juni 2024 Sonnenuntergang; Sonntag Sonnenuntergang bis Montag Sonnenuntergang]

3

3rd Day | Yom Shlishi | 10 June 2024 sunset to 11 June 2024 sunset; Monday sunset until Tuesday sunset

[3.Tag: 10.Juni 2024 Sonnenuntergang bis 11.Juni 2024 Sonnenuntergang; Montag Sonnenuntergang bis Dienstag Sonnenuntergang]

4

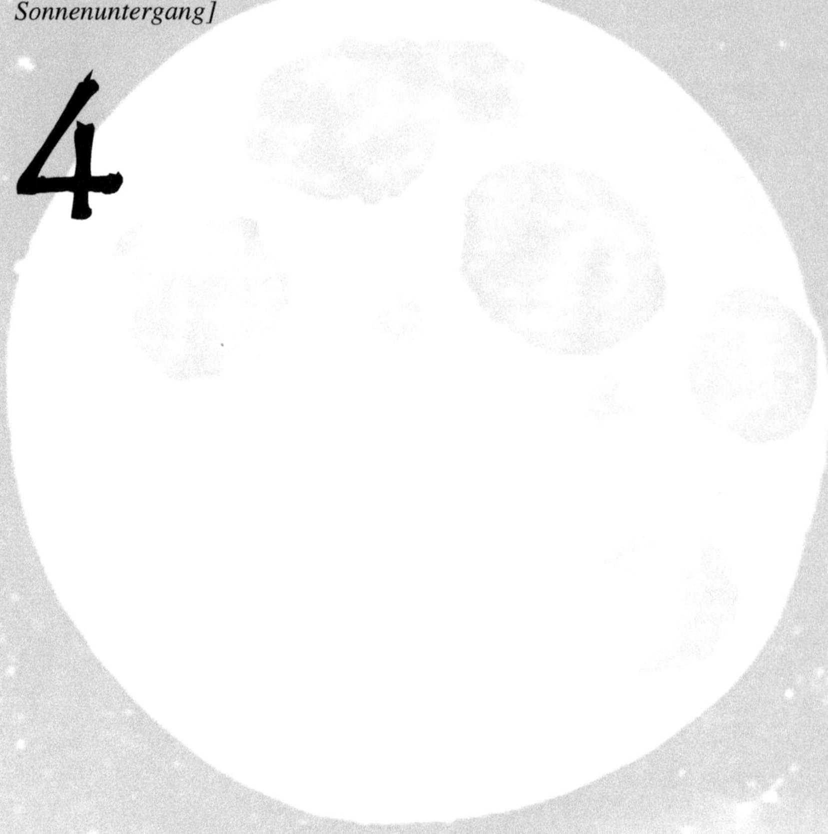

4th Day | Yom Revi´i | 11 June 2024 sunset to 12 June 2024 sunset; Tuesday sunset until Wednesday sunset

[4.Tag: 11.Juni 2024 Sonnenuntergang bis 12.Juni 2024 Sonnenuntergang; Dienstag Sonnenuntergang bis Mittwoch Sonnenuntergang]

5

5th Day | Yom Hamishi | 12 June 2024 sunset to 13 June 2024 sunset; Wednesday sunset until Thursday sunset

[5.Tag: 12.Juni 2024 Sonnenuntergang bis 13.Juni 2024 Sonnenuntergang; Mittwoch Sonnenuntergang bis Donnerstag Sonnenuntergang]

6

6th Day | Yom Shishi | 13 June 2024 sunset to 14 June 2024 sunset; Thursday sunset until Friday sunset

[6.Tag: 13.Juni 2024 Sonnenuntergang bis 14.Juni 2024 Sonnenuntergang; Donnerstag Sonnenuntergang bis Freitag Sonnenuntergang]

7

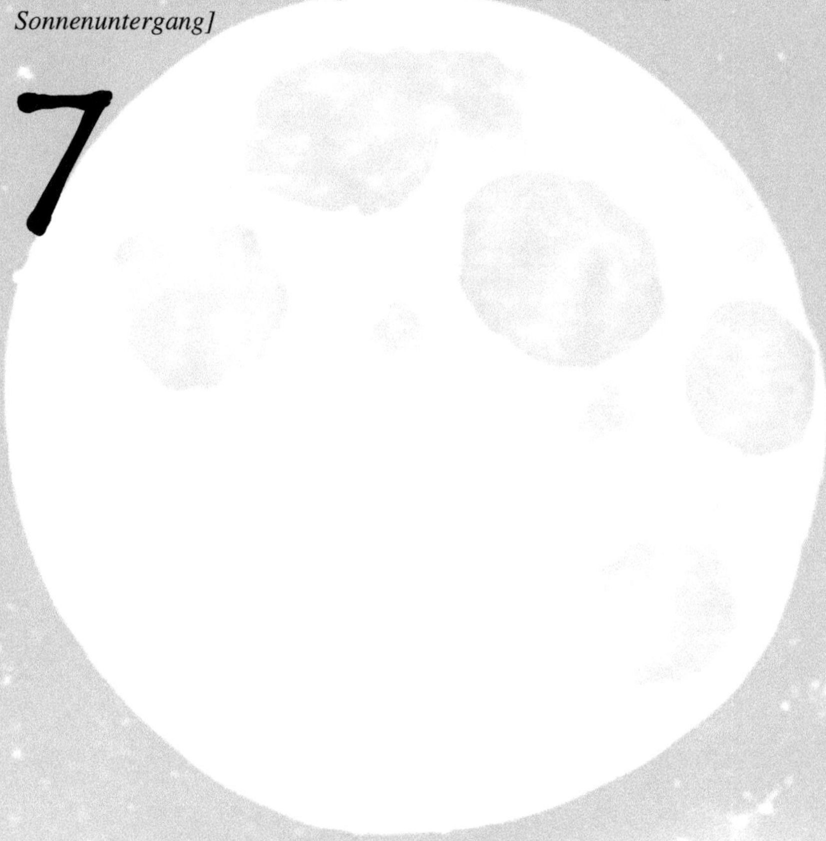

7th Day | Shabbat | 14 June 2024 sunset to 15 June 2024 sunset; Friday sunset until Saturday sunset

[7.Tag: 14.Juni 2024 Sonnenuntergang bis 15.Juni 2024 Sonnenuntergang; Freitag Sonnenuntergang bis Samstag Sonnenuntergang]

8

1st Day | Yom Rishon | 15 June 2024 sunset to 16 June 2024 sunset; Saturday sunset until Sunday sunset

[1.Tag: 15.Juni 2024 Sonnenuntergang bis 16.Juni 2024 Sonnenuntergang; Samstag Sonnenuntergang bis Sonntag Sonnenuntergang]

9

2nd Day | Yom Sheni | 16 June 2024 sunset to 17 June 2024 sunset; Sunday sunset until Monday sunset

[2.Tag: 16.Juni 2024 Sonnenuntergang bis 17.Juni 2024 Sonnenuntergang; Sonntag Sonnenuntergang bis Montag Sonnenuntergang]

10

3rd Day | Yom Shlishi | 17 June 2024 sunset to 18 June 2024 sunset; Monday sunset until Tuesday sunset

[3.Tag: 17.Juni 2024 Sonnenuntergang bis 18.Juni 2024 Sonnenuntergang; Montag Sonnenuntergang bis Dienstag Sonnenuntergang]

11

4th Day | Yom Revi ´i | 18 June 2024 sunset to 19 June 2024 sunset; Tuesday sunset until Wednesday sunset

[4.Tag: 18.Juni 2024 Sonnenuntergang bis 19.Juni 2024 Sonnenuntergang; Dienstag Sonnenuntergang bis Mittwoch Sonnenuntergang]

12

5th Day | Yom Hamishi | 19 June 2024 sunset to 20 June 2024 sunset; Wednesday sunset until Thursday sunset

[5.Tag: 19.Juni 2024 Sonnenuntergang bis 20.Juni 2024 Sonnenuntergang; Mittwoch Sonnenuntergang bis Donnerstag Sonnenuntergang]

13

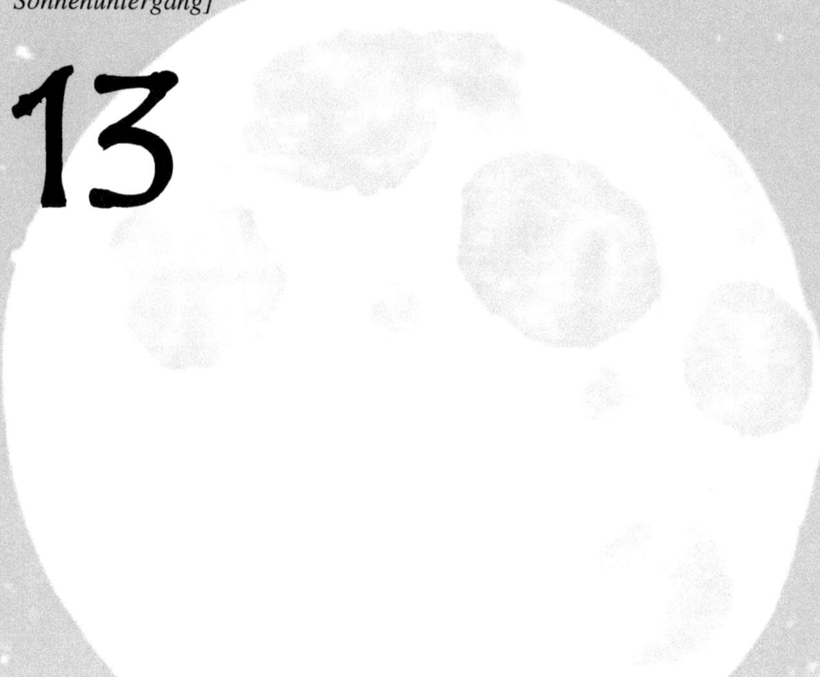

6th Day | Yom Shishi | Summer Solstice | 20 June 2024 sunset to 21 June 2024 sunset; Thursday sunset until Friday sunset: Full Moon

[6.Tag: Sommersonnenwende | 20.Juni 2024 Sonnenuntergang bis 21.Juni 2024 Sonnenuntergang; Donnerstag Sonnenuntergang bis Freitag Sonnenuntergang: Vollmond]

14 

7th Day | Shabbat | 21 June 2024 sunset to 22 June 2024 sunset; Friday sunset until Saturday sunset

[7.Tag: 21.Juni 2024 Sonnenuntergang bis 22.Juni 2024 Sonnenuntergang; Freitag Sonnenuntergang bis Samstag Sonnenuntergang]

15

1st Day | Yom Rishon | 22 June 2024 sunset to 23 June 2024 sunset; Saturday sunset until Sunday sunset

[1.Tag: 22.Juni 2024 Sonnenuntergang bis 23.Juni 2024 Sonnenuntergang; Samstag Sonnenuntergang bis Sonntag Sonnenuntergang]

16

2nd Day | Yom Sheni | Fast of 4th Month: For Aaron´s Sin of Making The Golden Calf (Exodus 32:1-5 | Ta ´anit 28 B) | 23 June 2024 sunset to 24 June 2024 sunset; Sunday sunset until Monday sunset

[2.Tag: Fasten des 4.Monats: Für Aarons Sünde, ein goldenes Kalb gemacht zu haben (Exodus/2.Mose 32,1-5 | Ta ´anit 28 B) | 23.Juni 2024 Sonnenuntergang bis 24.Juni 2024 Sonnenuntergang; Sonntag Sonnenuntergang bis Montag Sonnenuntergang]

17

3rd Day | Yom Shlishi | 24 June 2024 sunset to 25 June 2024 sunset; Monday sunset until Tuesday sunset

[3.Tag: 24.Juni 2024 Sonnenuntergang bis 25.Juni 2024 Sonnenuntergang; Montag Sonnenuntergang bis Dienstag Sonnenuntergang]

18

4th Day | Yom Revi ´i | 25 June 2024 sunset to 26 June 2024 sunset; Tuesday sunset until Wednesday sunset

[4.Tag: 25.Juni 2024 Sonnenuntergang bis 26.Juni 2024 Sonnenuntergang; Dienstag Sonnenuntergang bis Mittwoch Sonnenuntergang]

19

5th Day | Yom Hamishi | 26 June 2024 sunset to 27 June 2024 sunset; Wednesday sunset until Thursday sunset

[5.Tag: 26.Juni 2024 Sonnenuntergang bis 27.Juni 2024 Sonnenuntergang; Mittwoch Sonnenuntergang bis Donnerstag Sonnenuntergang]

20

6th Day | Yom Shishi | 27 June 2024 sunset to 28 June 2024 sunset; Thursday sunset until Friday sunset

[6.Tag: 27.Juni 2024 Sonnenuntergang bis 28.Juni 2024 Sonnenuntergang; Donnerstag Sonnenuntergang bis Freitag Sonnenuntergang]

21

7th Day | Shabbat | 28 June 2024 sunset to 29 June 2024 sunset; Friday sunset until Saturday sunset

[7.Tag: 28.Juni 2024 Sonnenuntergang bis 29.Juni 2024 Sonnenuntergang; Freitag Sonnenuntergang bis Samstag Sonnenuntergang]

22

1st Day | Yom Rishon | 29 June 2024 sunset to 30 June 2024 sunset; Saturday sunset until Sunday sunset

[1.Tag: 29.Juni 2024 Sonnenuntergang bis 30.Juni 2024 Sonnenuntergang; Samstag Sonnenuntergang bis Sonntag Sonnenuntergang]

23

2nd Day | Yom Sheni | 30 June 2024 sunset to 1 July 2024 sunset; Sunday sunset until Monday sunset

[2.Tag: 30.Juni 2024 Sonnenuntergang bis 1.Juli 2024 Sonnenuntergang; Sonntag Sonnenuntergang bis Montag Sonnenuntergang]

24

3rd Day | Yom Shlishi | 1 July 2024 sunset to 2 July 2024 sunset; Monday sunset until Tuesday sunset

[3.Tag: 1.Juli 2024 Sonnenuntergang bis 2.Juli 2024 Sonnenuntergang; Montag Sonnenuntergang bis Dienstag Sonnenuntergang]

25

4th Day | Yom Revi ´i | 2 July 2024 sunset to 3 July 2024 sunset; Tuesday sunset until Wednesday sunset

[4.Tag: 2.Juli 2024 Sonnenuntergang bis 3.Juli 2024 Sonnenuntergang; Dienstag Sonnenuntergang bis Mittwoch Sonnenuntergang]

26

5th Day | Yom Hamishi | 3 July 2024 sunset to 4 July 2024 sunset; Wednesday sunset until Thursday sunset

[5.Tag: 3.Juli 2024 Sonnenuntergang bis 4.Juli 2024 Sonnenuntergang; Mittwoch Sonnenuntergang bis Donnerstag Sonnenuntergang]

27

6th Day | Yom Shishi | 4 July 2024 sunset to 5 July 2024 sunset; Thursday sunset until Friday sunset: Old Light

[6.Tag: 4.Juli 2024 Sonnenuntergang bis 5.Juli 2024 Sonnenuntergang; Donnerstag Sonnenuntergang bis Freitag Sonnenuntergang: Altlicht]

28

7th Day | Shabbat | 5 July 2024 sunset to 6 July 2024 sunset; Friday sunset until Saturday sunset: New Moon

[7.Tag: 5.Juli 2024 Sonnenuntergang bis 6.Juli 2024 Sonnenuntergang; Freitag Sonnenuntergang bis Samstag Sonnenuntergang: Neumond]

29

1st Day | Yom Rishon | Rosh Chodesh: First Crescent Moon sunset, 7 July: New Light | 6 July 2024 sunset to 7 July 2024 sunset; Saturday sunset until Sunday sunset

[1.Tag: Rosh Chodesh: Erste sichtbare Mondsichel, Sonnenuntergang 7.Juli: Neulichti | 6.Juli 2024 Sonnenuntergang bis 7.Juli 2024 Sonnenuntergang; Samstag Sonnenuntergang bis Sonntag Sonnenuntergang]

30 

Month 5 / *Monat 5*

Notes | Notizen

2nd Day | Yom Sheni | Rosh Chodesh: New Light | 7 July 2024 sunset to 8 July 2024 sunset; Sunday sunset until Monday sunset

[2.Tag: Rosh Chodesh: Neulicht | 7.Juli 2024 Sonnenuntergang bis 8.Juli 2024 Sonnenuntergang; Sonntag Sonnenuntergang bis Montag Sonnenuntergang]

1 

3rd Day | Yom Shlishi | 8 July 2024 sunset to 9 July 2024 sunset; Monday sunset until Tuesday sunset

[3.Tag: 8.Juli 2024 Sonnenuntergang bis 9.Juli 2024 Sonnenuntergang; Montag Sonnenuntergang bis Dienstag Sonnenuntergang]

2

4th Day | Yom Revi ́i | 9 July 2024 sunset to 10 July 2024 sunset; Tuesday sunset until Wednesday sunset

[4.Tag: 9.Juli 2024 Sonnenuntergang bis 10.Juli 2024 Sonnenuntergang; Dienstag Sonnenuntergang bis Mittwoch Sonnenuntergang]

3

5th Day | Yom Hamishi | 10 July 2024 sunset to 11 July 2024 sunset; Wednesday sunset to Thursday sunset

[5.Tag: 10.Juli 2024 Sonnenuntergang bis 11.Juli 2024 Sonnenuntergang; Mittwoch Sonnenuntergang bis Donnerstag Sonnenuntergang]

4

6th Day | Yom Shishi | 11 July 2024 sunset to 12 July 2024 sunset; Thursday sunset until Friday sunset

[6.Tag: 11.Juli 2024 Sonnenuntergang bis 12.Juli 2024 Sonnenuntergang; Donnerstag Sonnenuntergang bis Freitag Sonnenuntergang]

5

7th Day | Shabbat | 12 July 2024 sunset to 13 July 2024 sunset; Friday sunset until Saturday sunset

[7.Tag: 12.Juli 2024 Sonnenuntergang bis 13.Juli 2024 Sonnenuntergang; Freitag Sonnenuntergang bis Samstag Sonnenuntergang]

6

1st Day | Yom Rishon | 13 July 2024 sunset to 14 July 2024 sunset; Saturday sunset until Sunday sunset

[1.Tag: 13.Juli 2024 Sonnenuntergang bis 14.Juli 2024 Sonnenuntergang; Samstag Sonnenuntergang bis Sonntag Sonnenuntergang]

7

2nd Day | Yom Sheni | 14 July 2024 sunset to 15 July 2024 sunset; Sunday sunset until Monday sunset

[2.Tag: 14.Juli 2024 Sonnenuntergang bis 15.Juli 2024 Sonnenuntergang; Sonntag Sonnenuntergang bis Montag Sonnenuntergang]

8

3rd Day | Yom Shlishi | Tisha B´Av: For the Sin of Believing the False Report of the 10 Spies (Numbers 13:25-14:12 | Zechariah 7:3-5, 8:19) | Ta´anit 26 B) | 15 July 2024 sunset to 16 July 2024 sunset; Monday sunset until Tuesday sunset

[3.Tag: Tisha B´Av: Für die Sünde, dem falschen Bericht der 10 Auskundschafter geglaubt zu haben (Numeri/4.Mose 13,25+14,12); Sacharja 7,3-5+8,19) | Ta´anit 26 B | 15.Juli 2024 Sonnenuntergang bis 16.Juli 2024 Sonnenuntergang; Montag Sonnenuntergang bis Dienstag Sonnenuntergang]

9

4th Day | Yom Revi´i | 16 July 2024 sunset to 17 July 2024 sunset; Tuesday sunset until Wednesday sunset

[4.Tag: 16.Juli 2024 Sonnenuntergang bis 17.Juli 2024 Sonnenuntergang; Dienstag Sonnenuntergang bis Mittwoch Sonnenuntergang]

10

5th Day | Yom Hamishi | 17 July 2024 sunset to 18 July 2024 sunset; Wednesday sunset until Thursday sunset

[5.Tag: 17.Juli 2024 Sonnenuntergang bis 18.Juli 2024 Sonnenuntergang; Mittwoch Sonnenuntergang bis Donnerstag Sonnenuntergang]

11

6th Day | Yom Shishi | 18 July 2024 sunset to 19 July 2024 sunset; Thursday sunset until Friday sunset

[6.Tag: 18.Juli 2024 Sonnenuntergang bis 19.Juli 2024 Sonnenuntergang; Donnerstag Sonnenuntergang bis Freitag Sonnenuntergang]

12

7th Day | Shabbat | 19 July 2024 sunset to 20 July 2024 sunset; Friday sunset until Saturday sunset

]7.Tag: 19.Juli 2024 Sonnenuntergang bis 20.Juli 2024 Sonnenuntergang; Freitag Sonnenuntergang bis Samstag Sonnenuntergang]

13

1st Day | Yom Rishon | 20 July 2024 sunset to 21 July 2024 sunset; Saturday sunset until Sunday sunset: Full Moon

[1.Tag: 20.Juli 2024 Sonnenuntergang bis 21.Juli 2024 Sonnenuntergang; Samstag Sonnenuntergang bis Sonntag Sonnenuntergang: Vollmond]

14

2nd Day | Yom Sheni | Tu B´Av: Yahuah speaks to Moses (Deuteronomy 2:1-25), Ta´anit 30 B | 21 July 2024 sunset to 22 July 2024 sunset; Sunday sunset until Monday sunset

[2.Tag: Tu B´Av: Yahuah spricht zu Moses (Deuteronomium/5.Mose 2,1-25), Ta ´anit 30 B | 21.Juli 2024 Sonnenuntergang bis 22.Juli 2024 Sonnenuntergang; Sonntag Sonnenuntergang bis Montag Sonnenuntergang]

15

3rd Day | Yom Shlishi | 22 July 2024 sunset to 23 July 2024 sunset; Monday sunset until Tuesday sunset

[3.Tag: 22.Juli 2024 Sonnenuntergang bis 23.Juli 2024 Sonnenuntergang; Montag Sonnenuntergang bis Dienstag Sonnenuntergang]

16

4th Day | Yom Revi´i | 23 July 2024 sunset to <u>24 July</u> 2024 sunset; Tuesday sunset until <u>Wednesday</u> sunset

[4.Tag: 23.Juli 2024 Sonnenuntergang bis <u>24.Juli</u> 2024 Sonnenuntergang; Dienstag Sonnenuntergang bis <u>Mittwoch</u> Sonnenuntergang]

17

5th Day ¦ Yom Hamishi | 24 July 2024 sunset to 25 July 2024 sunset; Wednesday sunset until Thursday sunset

[5.Tag: 25.Juli 2024 Sonnenuntergang bis 26.Juli 2024 Sonnenuntergang; Mittwoch Sonnenuntergang bis Donnerstag Sonnenuntergang]

18

6th Day | Yom Shishi | 25 July 2024 sunset to <u>26 July</u> 2024 sunset; Thursday sunset until <u>Friday</u> sunset

[6.Tag: 25.Juli 2024 Sonnenuntergang bis <u>26.Juli</u> 2024 Sonnenuntergang; Donnerstag Sonnenuntergang bis <u>Freitag</u> Sonnenuntergang]

19

7th Day | Shabbat | 26 July 2024 sunset to 27 July 2024 sunset; Friday sunset until Saturday sunset

[7.Tag: 26.Juli 2024 Sonnenuntergang bis 27.Juli 2024 Sonnenuntergang; Freitag Sonnenuntergang bis Samstag Sonnenuntergang]

20

1st Day | Yom Rishon | 27 July 2024 sunset to 28 July 2024 sunset; Saturday sunset until Sunday sunset

[1.Tag: 27.Juli 2024 Sonnenuntergang bis 28.Juli 2024 Sonnenuntergang; Samstag Sonnenuntergang bis Sonntag Sonnenuntergang]

21

2nd Day | Yom Sheni | 28 July 2024 sunset to 29 July 2024 sunset; Sunday sunset until Monday sunset

[2.Tag: 28.Juli 2024 Sonnenuntergang bis 29.Juli 2024 Sonnenuntergang; Sonntag Sonnenuntergang bis Montag Sonnenuntergang]

22

3rd Day | Yom Shlishi | 29 July 2024 sunset to 30 July 2024 sunset; Monday sunset until Tuesday sunset

[3.Tag: 29.Juli 2024 Sonnenuntergang bis 30.Juli 2024 Sonnenuntergang; Montag Sonnenuntergang bis Dienstag Sonnenuntergang]

23

4th Day | Yom Revi´i | 30 July 2024 sunset to 31 July 2024 sunset; Tuesday sunset until Wednesday sunset

[4.Tag: 30.Juli 2024 Sonnenuntergang bis 31.Juli 2024 Sonnenuntergang; Dienstag Sonnenuntergang bis Mittwoch Sonnenuntergang]

24

5th Day | Yom Hamishi | 31 July 2024 sunset to 1 August 2024 sunset; Wednesday sunset until Thursday sunset

[5.Tag: 31.Juli 2024 Sonnenuntergang bis 1.August 2024 Sonnenuntergang; Mittwoch Sonnenuntergang bis Donnerstag Sonnenuntergang]

25

6th Day | Yom Shishi | 1 August 2024 sunset to 2 August 2024 sunset; Thursday sunset until Friday sunset

[6.Tag: 1.August 2024 Sonnenuntergang bis 2.August 2024 Sonnenuntergang; Donnerstag Sonnenuntergang bis Freitag Sonnenuntergang]

26

7th Day | Shabbat | 2 August 2024 sunset to 3 August 2024 sunset; Friday sunset until Saturday sunset: Old Light

[7.Tag: 2.August 2024 Sonnenuntergang bis 3.August 2024 Sonnenuntergang; Freitag Sonnenuntergang bis Samstag Sonnenuntergang: Altlicht]

27

1st Day | Yom Rishon | 3 August 2024 sunset to 4 August 2024 sunset; Saturday sunset until Sunday sunset: New Moon

[1.Tag: 3.August 2024 Sonnenuntergang bis 4.August 2024 Sonnenuntergang; Samstag Sonnenuntergang bis Sonntag Sonnenuntergang: Neumond]

28

2nd Day | Yom Sheni | Rosh Chodesh: First Crescent Moon sunset, 5 August: New Light | 4 August 2024 sunset to <u>5 August</u> 2024 sunset; Sunday sunset until <u>Monday</u> sunset

[2.Tag: Rosh Chodesh: Erste sichtbare Mondsichel, Sonnenuntergang 5.August: Neulicht | 4.August 2024 Sonnenuntergang bis 5.August 2024 Sonnenuntergang; Sonntag Sonnenuntergang bis Montag Sonnenuntergang]

29

Notes | Notizen

Month 6 | *Monat 6*

Notes | Notizen

3rd Day | Yom Shlishi | Rosh Chodesh: New Light | 5 August 2024 sunset to 6 August 2024 sunset; Monday sunset until Tuesday sunset

[3.Tag: Rosh Chodesh: Neulicht | 5.August 2024 Sonnenuntergang bis 6.August 2024 Sonnenuntergang; Montag Sonnenuntergang bis Dienstag Sonnenuntergang]

1

4th Day | Yom Revi´i | 6 August 2024 sunset to 7 August 2024 sunset; Tuesday sunset until Wednesday sunset

[4.Tag: 6.August 2024 Sonnenuntergang bis 7.August 2024 Sonnenuntergang; Dienstag Sonnenuntergang bis Mittwoch Sonnenuntergang]

2

5th Day | Yom Hamishi | 7 August 2024 sunset to 8 August 2024 sunset; Wednesday sunset until Thursday sunset

[5.Tag: 7.August 2024 Sonnenuntergang bis 8.August 2024 Sonnenuntergang; Mittwoch Sonnenuntergang bis Donnerstag Sonnenuntergang]

3

6th Day | Yom Shishi | 8 August 2024 sunset to 9 August 2024 sunset; Thursday sunset until Friday sunset

[6.Tag: 8.August 2024 Sonnenuntergang bis 9.August 2024 Sonnenuntergang; Donnerstag Sonnenuntergang bis Freitag Sonnenuntergang]

4

7th Day | Shabbat | 9 August 2024 sunset to 10 August 2024 sunset; Friday sunset until Saturday sunset

[7.Tag: 9.August 2024 Sonnenuntergang bis 10.August 2024 Sonnenuntergang; Freitag Sonnenuntergang bis Samstag Sonnenuntergang]

5

1st Day | Yom Rishon | 10 August 2024 sunset to 11 August 2024 sunset; Saturday sunset until Sunday sunset

[1.Tag: 10.August 2024 Sonnenuntergang bis 11.August 2024 Sonnenuntergang; Samstag Sonnenuntergang bis Sonntag Sonnenuntergang]

6

2nd Day | Yom Sheni | 11 August 2024 sunset to 12 August 2024 sunset; Sunday sunset until Monday sunset

[2.Tag: 11.August 2024 Sonnenuntergang bis 12.August 2024 Sonnenuntergang; Sonntag Sonnenuntergang bis Montag Sonnenuntergang]

7

3rd Day | Yom Shlishi | 12 August 2024 sunset to 13 August 2024 sunset; Monday sunset until Tuesday sunset

[3.Tag: 12.August 2024 Sonnenuntergang bis 13.August 2024 Sonnenuntergang; Montag Sonnenuntergang bis Dienstag Sonnenuntergang]

8

4th Day | Yom Revi´i | 13 August 2024 sunset to 14 August 2024 sunset; Tuesday sunset until Wednesday sunset

[4.Tag: 13.August 2024 Sonnenuntergang bis 14.August 2024 Sonnenuntergang; Dienstag Sonnenuntergang bis Mittwoch Sonnenuntergang]

9

5th Day | Yom Hamishi | 14 August 2024 sunset to 15 August 2024 sunset; Wednesday sunset until Thursday sunset

[5.Tag: 14.August 2024 Sonnenuntergang bis 15.August 2024 Sonnenuntergang; Mittwoch Sonnenuntergang bis Donnerstag Sonnenuntergang]

10

6th Day | Yom Shishi | 15 August 2024 sunset to 16 August 2024 sunset; Thursday sunset until Friday sunset

[6.Tag: 15.August 2024 Sonnenuntergang bis 16.August 2024 Sonnenuntergang; Donnerstag Sonnenuntergang bis Freitag Sonnenuntergang]

11

7th Day | Shabbat | 16 August 2024 sunset to 17 August 2024 sunset; Friday sunset until Saturday sunset

[7.Tag: 16.August 2024 Sonnenuntergang bis 17.August 2024 Sonnenuntergang; Freitag Sonnenuntergang bis Samstag Sonnenuntergang]

12

1st Day | Yom Rishon | 17 August 2024 sunset to 18 August 2024 sunset; Saturday sunset until Sunday sunset

[1.Tag: 17.August 2024 Sonnenuntergang bis 18.August 2024 Sonnenuntergang; Samstag Sonnenuntergang bis Sonntag Sonnenuntergang]

13

2nd Day | Yom Sheni | 18 August 2024 sunset to 19 August 2024 sunset; Sunday sunset until Monday sunset: Full Moon

[2.Tag: 18.August 2024 Sonnenuntergang bis 19.August 2024 Sonnenuntergang; Sonntag Sonnenuntergang bis Montag Sonnenuntergang: Vollmond]

14

3rd Day | Yom Shlishi | 19 August 2024 sunset to 20 August 2024 sunset; Monday sunset until Tuesday sunset

[3.Tag: 19.August 2024 Sonnenuntergang bis 20.August 2024 Sonnenuntergang; Montag Sonnenuntergang bis Dienstag Sonnenuntergang]

15

4th Day | Yom Revi´i | 20 August 2024 sunset to 21 August 2024 sunset; Tuesday sunset until Wednesday sunset

[4.Tag: 20.August 2024 Sonnenuntergang bis 21.August 2024 Sonnenuntergang; Dienstag Sonnenuntergang bis Mittwoch Sonnenuntergang]

16

5th Day | Yom Hamishi | 21 August 2024 sunset to 22 August 2024 sunset; Wednesday sunset until Thursday sunset

[5.Tag: 21.August 2024 Sonnenuntergang bis 22.August 2024 Sonnenuntergang; Mittwoch Sonnenuntergang bis Donnerstag Sonnenuntergang]

17

6th Day | Yom Shishi | 22 August 2024 sunset to 23 August 2024 sunset; Thursday sunset until Friday sunset

[6.Tag: 22.August 2024 Sonnenuntergang bis 23.August 2024 Sonnenuntergang; Donnerstag Sonnenuntergang bis Freitag Sonnenuntergang]

18

7th Day | Shabbat | 23 August 2024 sunset to 24 August 2024 sunset; Friday sunset until Saturday sunset

[7.Tag: 23.August 2024 Sonnenuntergang bis 24.August 2024 Sonnenuntergang; Freitag Sonnenuntergang bis Samstag Sonnenuntergang]

19

1st Day | Yom Rishon | 24 August 2024 sunset to 25 August 2024 sunset; Saturday sunset until Sunday sunset

[1.Tag: 24.August 2024 Sonnenuntergang bis 25.August 2024 Sonnenuntergang, Samstag Sonnenuntergang bis Sonntag Sonnenuntergang]

20

2nd Day | Yom Sheni | 25 August 2024 sunset to 26 August 2024 sunset; Sunday sunset until Monday sunset

[2.Tag: 25.August 2024 Sonnenuntergang bis 26.August 2024 Sonnenuntergang; Sonntag Sonnenuntergang bis Montag Sonnenuntergang]

21

3rd Day | Yom Shlishi | 26 August 2024 sunset to 27 August 2024 sunset; Monday sunset until Tuesday sunset

[3.Tag: 26.August 2024 Sonnenuntergang bis 27.August 2024 Sonnenuntergang; Montag Sonnenuntergang bis Dienstag Sonnenuntergang]

22

4th Day | Yom Revi´i | 27 August 2024 sunset to 28 August 2024 sunset; Tuesday sunset until Wednesday sunset

[4.Tag: 27.August 2024 Sonnenuntergang bis 28.August 2024 Sonnenuntergang; Dienstag Sonnenuntergang bis Mittwoch Sonnenuntergang]

23

5th Day | Yom Hamishi | 28 August 2024 sunset to 29 August 2024 sunset; Wednesday sunset until Thursday sunset

[5.Tag: 28.August 2024 Sonnenuntergang bis 29.August 2024 Sonnenuntergang; Mittwoch Sonnenuntergang bis Donnerstag Sonnenuntergang]

24

6th Day | Yom Shishi | 29 August 2024 sunset to 30 August 2024 sunset; Thursday sunset until Friday sunset

[6.Tag: 29.August 2024 Sonnenuntergang bis 30.August 2024 Sonnenuntergang; Donnerstag Sonnenuntergang bis Freitag Sonnenuntergang]

25

7th Day | Shabbat | 30 August 2024 sunset to 31 August 2024 sunset; Friday sunset until Saturday sunset

[7.Tag: 30.August 2024 Sonnenuntergang bis 31.August 2024 Sonnenuntergang; Freitag Sonnenuntergang bis Samstag Sonnenuntergang]

26

1st Day | Yom Rishon | 31 August 2024 sunset to 1 September 2024 sunset; Saturday sunset until Sunday sunset

[1.Tag: 31.August 2024 Sonnenuntergang bis 1.September 2024 Sonnenuntergang; Samstag Sonnenuntergang bis Sonntag Sonnenuntergang]

27

2nd Day | Yom Sheni | 1 September 2024 sunset to 2 September 2024 sunset; Sunday sunset until Monday sunset: Old Light

[2.Tag: 1.September 2024 Sonnenuntergang bis 2.September 2024 Sonnenuntergang; Sonntag Sonnenuntergang bis Montag Sonnenuntergang: Altlicht]

28

3rd Day | Yom Shlishi | 2 September 2024 sunset to 3 September 2024 sunset; Monday sunset until Tuesday sunset: New Moon

[3.Tag: 2.September 2024 Sonnenuntergang bis 3.September 2024 Sonnenuntergang; Montag Sonnenuntergang bis Dienstag Sonnenuntergang: Neumond]

29

4th Day | Yom Revi´i | Rosh Chodesh: First Crescent Moon sunset, 4 September: New Light | 3 September 2024 sunset to 4 September 2024 sunset; Tuesday sunset until Wednesday sunset

[4.Tag: Rosh Chodesh: Erste sichtbare Mondsichel, Sonnenuntergang 4.September: Neulicht | 3.September 2024 Sonnenuntergang bis 4.September 2024 Sonnenuntergang; Dienstag Sonnenuntergang bis Mittwoch Sonnenuntergang:]

30 

Month 7 | *Monat 7*

Notes | Notizen

5th Day | Yom Hamishi | Rosh Chodesh: New Light | Rosh Hashanah: Jewish/Civil New Year | Yom Teruah: Day of Trumpets | Birthday of Yahushua Ha-Mashiach (according to a sign in Revelation 12:1-2) | Begin „10 Days of Awe“, Day 1| 4 September 2024 sunset to 5 September 2024 sunset; Wednesday sunset until Thursday sunset

[5.Tag: Rosh Chodesh: Neulicht | Rosh Hashana: Jüdisches/Bürgerliches Neues Jahr | Yom Teruah: Posaunenfest | Geburtstag von Yahushua Ha-Mashiach (gemäß eines Zeichens in Offenbarung 12,1-2) | Beginn „10 Tage der Umkehr“, Tag 1 | 4.September Sonnenuntergang 2024 bis 5.September 2024 Sonnenuntergang; Mittwoch Sonnenuntergang bis Donnerstag Sonnenuntergang]

1 

6th Day | Yom Shishi | Yamim Noraim: Ten Days of Awe, Day 2 | 5 September 2024 sunset to 6 September 2024 sunset; Thursday sunset until Friday sunset

[6.Tag: Yamim Noraim: Zehn Tage der Umkehr, Tag 2 | 5.September 2024 Sonnenuntergang bis 6.September 2024 Sonnenuntergang; Donnerstag Sonnenuntergang bis Freitag Sonnenuntergang]

2

7th Day | Shabbat | | Yamim Noraim: Ten Days of Awe, Day 3 | 6 September 2024 sunset to 7 September 2024 sunset; Friday sunset until Saturday sunset

[7.Tag: Yamim Noraim: Zehn Tage der Umkehr, Tag 3 | 6.September 2024 Sonnenuntergang bis 7.September 2024 Sonnenuntergang; Freitag Sonnenuntergang bis Samstag Sonnenuntergang]

3

1st Day | Yom Rishon | Tzom Gedaliah: Fast of Gedaliah | Yamim Noraim: Ten Days of Awe, Day 4 (Usually observed Day 3 of Month 7) | 7 September 2024 sunset to 8 September 2024 sunset; Saturday sunset until Sunday sunset

[1.Tag: Tzom Gedaliah: Gedaliah-Fasten | Yamim Noraim: Zehn Tage der Umkehr, Tag 4 (gewöhnlich am 3.Tag des 7.Monats) | 7.September 2024 Sonnenuntergang bis 8.September 2024 Sonnenuntergang; Samstag Sonnenuntergang bis Sonntag Sonnenuntergang]

4

2nd Day | Yom Sheni | Yamim Noraim: Ten Days of Awe, Day 5 | 8 September 2024 sunset to <u>9 September</u> 2024 sunset; Sunday sunset until <u>Monday</u> sunset

[2.Tag: Yamim Noraim: Zehn Tage der Umkehr, Tag 5 | 8.September 2024 Sonnenuntergang bis <u>9.September</u> 2024 Sonnenuntergang; Sonntag Sonnenuntergang bis <u>Montag</u> Sonnenuntergang]

5

3rd Day | Yom Shlishi | Yamim Noraim: Ten Days of Awe, Day 6 | 9 September 2024 sunset to 10 September 2024 sunset; Monday sunset until Tuesday sunset

[3.Tag: Yamim Noraim: Zehn Tage der Umkehr, Tag 6 | 9.September 2024 Sonnenuntergang bis 10.September 2024 Sonnenuntergang; Montag Sonnenuntergang bis Dienstag Sonnenuntergang]

6

4th Day | Yom Revi´i | Yamim Noraim: Ten Days of Awe, Day 7 | 10 September 2024 sunset to 11 September 2024 sunset; Tuesday sunset until Wednesday sunset

[4.Tag: Yamim Noraim: Zehn Tage der Umkehr, Tag 7 | 10.September 2024 Sonnenuntergang bis 11.September 2024 Sonnenuntergang; Dienstag Sonnenuntergang bis Mittwoch Sonnenuntergang]

7

5th Day | Yom Hamishi | Yamim Noraim: Ten Days of Awe, Day 8 | 11 September 2024 sunset to 12 September 2023 sunset; Wednesday sunset until Thursday sunset

[5.Tag: Yamim Noraim: Zehn Tage der Umkehr, Tag 8 | 11.September 2024 Sonnenuntergang bis 12.September 2024 Sonnenuntergang; Mittwoch Sonnenuntergang bis Donnerstag Sonnenuntergang]

8

6th Day | Yom Shishi | Yamim Noraim: Ten Days of Awe, Day 9 | 12 September 2024 sunset to 13 September 2024 sunset; Thursday sunset until Friday sunset

[6.Tag: Yamim Noraim: Zehn Tage der Umkehr, Tag 9 | 12.September 2024 Sonnenuntergang bis 13.September 2024 Sonnenuntergang; Donnerstag Sonnenuntergang bis Freitag Sonnenuntergang]

9

7th Day | Shabbat | Yom Kippur: Day of Atonement | Yamim Noraim: Ten Days of Awe, Day 10 | 13 September 2024 sunset to 14 September 2024 sunset; Friday sunset until Saturday sunset

[7.Tag: Yom Kippur: Versöhnungstag | Yamim Noraim: Zehn Tage der Umkehr, Tag 10 | 13.September 2024 Sonnenuntergang bis 14.September 2024 Sonnenuntergang; Freitag Sonnenuntergang bis Samstag Sonnenuntergang]

10

1st Day | Yom Rishon | 14 September 2024 sunset to 15 September 2024 sunset; Saturday sunset until Sunday sunset

[1.Tag: 14.September 2024 Sonnenuntergang bis 15.September 2024 Sonnenuntergang; Samstag Sonnenuntergang bis Sonntag Sonnenuntergang]

11

2nd Day | Yom Sheni | 15 September 2024 sunset to 16 September 2024 sunset; Sunday sunset until Monday sunset

[2.Tag: 15.September 2024 Sonnenuntergang bis 16.September 2024 Sonnenuntergang; Sonntag Sonnenuntergang bis Montag Sonnenuntergang]

12

3rd Day | Yom Shlishi | 16 September 2024 sunset to 17 September 2024 sunset; Monday sunset until Tuesday sunset

[3.Tag: 16.September 2024 Sonnenuntergang bis 17.September 2024 Sonnenuntergang; Montag Sonnenuntergang bis Dienstag Sonnenuntergang]

13

4th Day | Yom Revi´i | Partial Lunar Eclipse/Full Moon | Eve of Sukkot | 17 September 2024 sunset to 18 September 2024 sunset; Tuesday sunset until Wednesday sunset

[4.Tag: Partielle Mondfinsternis/Vollmond | Abend vor Sukkot | 17.September 2024 Sonnenuntergang bis 18.September 2024 Sonnenuntergang; Dienstag Sonnenuntergang bis Mittwoch Sonnenuntergang]

14  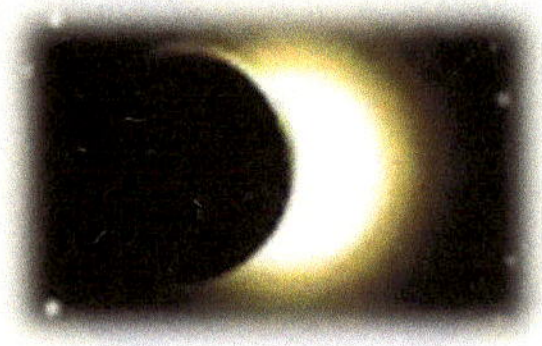

5th Day | Yom Hamishi | Sukkot: Tabernacles Day 1/Sukkot Day 1 | 18 September 2024 sunset to 19 September 2024 sunset; Wednesday sunset until Thursday sunset

[5.Tag: Sukkot: Laubhüttenfest Tag 1/Sukkot Tag 1 | 18.September 2024 Sonnenuntergang bis 19.September 2024 Sonnenuntergang; Mittwoch Sonnenuntergang bis Donnerstag Sonnenuntergang]

15

6th Day | Yom Shishi | Sukkot: Tabernacles Day 2/Sukkot Day 2 | 19 September 2024 sunset to 20 September 2024 sunset; Thursday sunset until Friday sunset

[6.Tag: Sukkot: Laubhüttenfest Tag 2/Sukkot Tag 2 | 19.September 2024 Sonnenuntergang bis 20.September 2024 Sonnenuntergang; Donnerstag Sonnenuntergang bis Freitag Sonnenuntergang]

16

7th Day | Shabbat | Sukkot: Tabernacles Day 3/Sukkot Day 3 | 20 September 2024 sunset to 21 September 2024 sunset; Friday sunset until Saturday sunset

[7.Tag: Sukkot: Laubhüttenfest Tag 3/Sukkot Tag 3 | 20.September 2024 Sonnenuntergang bis 21.September 2024 Sonnenuntergang; Freitag Sonnenuntergang bis Samstag Sonnenuntergang]

17

1st Day | Yom Rishon | Fall Equinox (22 Sep, 12h:44m UT) | Sukkot: Tabernacles Day 4/Sukkot Day 4 | 21 September 2024 sunset to 22 September 2024 sunset; Saturday sunset until Sunday sunset

[1.Tag: Herbstäquinoktium: Tag-und-Nacht-Gleiche (22.Sep, 12h:44m UT) | Sukkot: Laubhüttenfest Tag 4/Sukkot Tag 4 | 21.September 2024 Sonnenuntergang bis 22.September 2024 Sonnenuntergang; Samstag Sonnenuntergang bis Sonntag Sonnenuntergang]

18

2nd Day | Yom Sheni | Sukkot: Tabernacles Day 5/Sukkot Day 5 | 22 September 2024 sunset to 23 September 2024 sunset; Sunday sunset until Monday sunset

[2.Tag: Sukkot: Laubhüttenfest Tag 5/Sukkot Tag 5 | 22.September 2024 Sonnenuntergang bis 23.September 2024 Sonnenuntergang; Sonntag Sonnenuntergang bis Montag Sonnenuntergang]

19

3rd Day | Yom Shlishi | Sukkot: Tabernacles Day 6/Sukkot Day 6 | 23 September 2024 sunset to 24 September 2024 sunset; Monday sunset until Tuesday sunset

[3.Tag: Sukkot: Laubhüttenfest Tag 6/Sukkot Tag 6 | 23.September 2024 Sonnenuntergang bis 24.September 2024 Sonnenuntergang; Montag Sonnenuntergang bis Dienstag Sonnenuntergang]

20

4th Day | Yom Revi´i | Hoshana Rabbah: The Great Hoshana | Sukkot: Tabernacles Day 7/Sukkot Day 7 | 24 September 2024 sunset to 25 September 2024 sunset; Tuesday sunset until Wednesday sunset

[4.Tag: Hoshana Rabba: Großes Hoshana | Sukkot: Laubhüttenfest Tag 7/Sukkot Tag 7 | 24.September 2024 Sonnenuntergang bis 25.September 2024 Sonnenuntergang; Dienstag Sonnenuntergang bis Mittwoch Sonnenuntergang]

21

5th Day | Yom Hamishi | Shemini Atzeret: 8th Day Solemn Assembly (Leviticus 23:36B | Numbers 29:35) | 25 September 2024 sunset to <u>26 September</u> 2024 sunset; Wednesday sunset until <u>Thursday</u> sunset

[5.Tag: Shemini Atzeret: 8ter Tag Feierliche Versammlung [Levitikus 23,36B | Numeri 29,35) | 25.September 2024 Sonnenuntergang bis <u>26 September</u> 2024 Sonnenuntergang; Mittwoch Sonnenuntergang bis <u>Donnerstag</u> Sonnenuntergang]

22

6th Day | Yom Shlishi | Simchat Torah: Rejoicing in the Good Things Yahuah has done (2 Chronicles 7:10) | 26 September 2024 sunset to 27 September 2024 sunset; Thursday sunset until Friday sunset

[6.Tag: Simchat Tora: Freude über die Tora/Freude über die Guten Dinge, die Yahuah getan hat (2.Chronik 7,10) | 26.September 2024 Sonnenuntergang bis 27.September 2024 Sonnenuntergang; Donnerstag Sonnenuntergang bis Freitag Sonnenuntergang]

23

7th Day | Shabbat | 27 September 2024 sunset to 28 September 2024 sunset; Friday sunset until Saturday sunset

[7.Tag: 27.September 2024 Sonnenuntergang bis 28.September 2024; Freitag Sonnenuntergang bis Samstag Sonnenuntergang]

24

1st Day | Yom Rishon | Israel Set Apart: Israel turn to Yahuah in their Distress (Nehemiah 9:1-38), Usually Observed Day 24 of Month 7 | 28 September 2024 sunset to 29 September 2024 sunset; Saturday sunset until Sunday sunset

[1.Tag: Absonderung Israel: Israel wendet sich in seiner Not Yahuah zu (Nehemia 9,1-38, gewöhnlich am 24.Tag des 7.Monats | 28.September 2024 Sonnenuntergang bis 29.September 2024 Sonnenuntergang; Samstag Sonnenuntergang bis Sonntag Sonnenuntergang]

25

2nd Day | Yom Sheni | 29 September 2024 sunset to 30 September 2024 sunset; Sunday sunset until Monday sunset

[2.Tag: 29.September 2024 Sonnenuntergang bis 30.September 2024 Sonnenuntergang; Sonntag Sonnenuntergang bis Montag Sonnenuntergang]

26

3rd Day | Yom Shlishi | 30 September 2024 sunset to 1 October 2024 sunset; Monday sunset until Tuesday sunset

[3.Tag: 30.September 2024 Sonnenuntergang bis 1.Oktober 2024 Sonnenuntergang; Montag Sonnenuntergang bis Dienstag Sonnenuntergang]

27

4th Day | Yom Revi´i | 1 October 2024 sunset to 2 October 2024 sunset; Tuesday sunset until Wednesday sunset: Old Light

[4.Tag: 1.Oktober 2024 Sonnenuntergang bis 2.Oktober 2024 Sonnenuntergang; Dienstag Sonnenuntergang bis Mittwoch Sonnenuntergang: Altlicht]

28

5th Day | Yom Hamishi | Annular Solar Eclipse/New Moon | 2 October 2024 sunset to 3 October 2024 sunset; Wednesday sunset until Thursday sunset

[5.Tag: Ringförmige Sonnenfinsternis/Neumond | 2.Oktober 2024 Sonnenuntergang bis 3.Oktober 2024 Sonnenuntergang; Mittwoch Sonnenuntergang bis Donnerstag Sonnenuntergang]

29  

6th Day | Yom Shishi | Rosh Chodesh: First Crescent Moon sunset, 4 Oct: New Light | 3 October 2024 sunset to 4 October 2024 sunset; Thursday sunset until Friday sunset

[6.Tag: Rosh Chodesh: Erste sichtbare Mondsichel, Sonnenuntergang 4.Okt: Neulicht | 3.Oktober 2024 Sonnenuntergang bis 4.Oktober 2024 Sonnenuntergang; Donnerstag Sonnenuntergang bis Freitag Sonnenuntergang]

30

Month 8 | *Monat 8*

Notes | Notizen

7th Day | Shabbat/Shabbat Rosh Chodesh | Rosh Chodesh: New Light | 4 October 2024 sunset to 5 October 2024 sunset; Friday sunset until Saturday sunset

[7.Tag: Rosh Chodesh: Neulicht | 4.Oktober 2024 Sonnenuntergang bis 5.Oktober 2024 Sonnenuntergang; Freitag Sonnenuntergang bis Samstag Sonnenuntergang]

1 

1st Day | Yom Rishon | 5 October 2024 sunset to 6 October 2024 sunset; Saturday sunset until Sunday sunset

[1.Tag: 5.Oktober 2024 Sonnenuntergang bis 6.Oktober 2024 Sonnenuntergang; Samstag Sonnenuntergang bis Sonntag Sonnenuntergang]

2

2nd Day | Yom Sheni | 6 October 2024 sunset to 7 October 2024 sunset; Sunday sunset until Monday sunset

[2.Tag: 6.Oktober 2024 Sonnenuntergang bis 7.Oktober 2024 Sonnenuntergang; Sonntag Sonnenuntergang bis Montag Sonnenuntergang]

3

3rd Day | Yom Shlishi | 7 October 2024 sunset to 8 October 2024 sunset; Monday sunset until Tuesday sunset

[3.Tag: 7.Oktober 2024 Sonnenuntergang bis 8.Oktober 2024 Sonnenuntergang; Montag Sonnenuntergang bis Dienstag Sonnenuntergang]

4

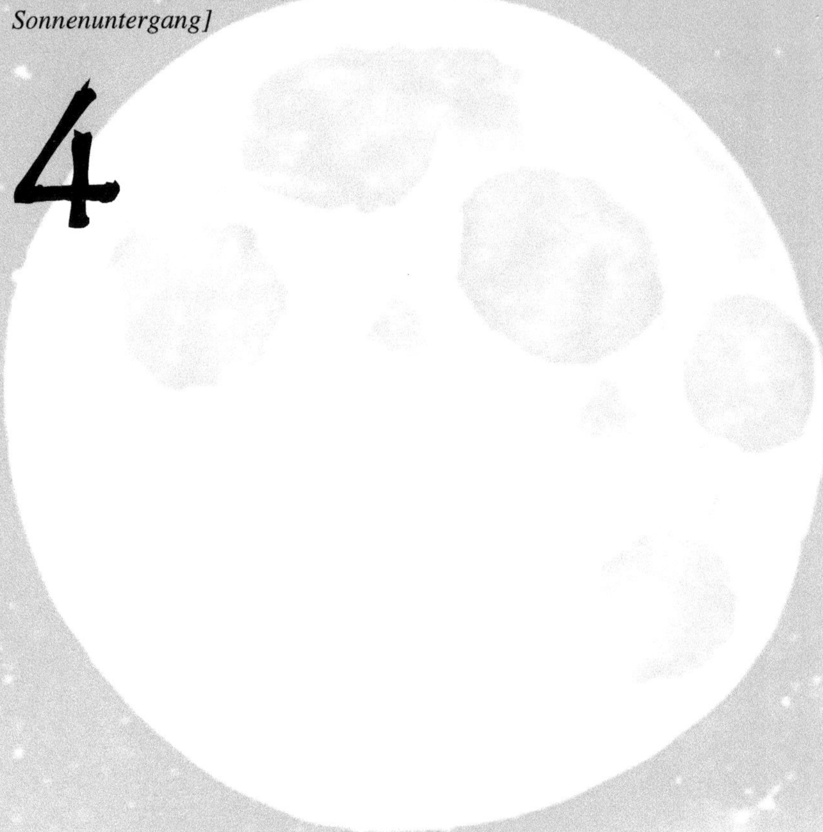

4th Day | Yom Revi´i | 8 October 2024 sunset to 9 October 2024 sunset; Tuesday sunset until Wednesday sunset

[4.Tag: 8.Oktober 2024 Sonnenuntergang bis 9.Oktober 2024 Sonnenuntergang; Dienstag Sonnenuntergang bis Mittwoch Sonnenuntergang]

5

5th Day | Yom Hamishi | Fast of Zedekiah (2 Kings 25:6-9 | Jeremiah 39:5-7 | Jeremiah 52:9-11); Megillat Ta´anit | 9 October 2024 sunset to 10 October 2024; Wednesday sunset until Thursday sunset

[5.Tag: Zedekia-Fasten (2.Könige 25,6-9 | Jeremia 39,5-7 | Jeremia 52, 9-11); Megillat Ta´anit | 9.Oktober 2024 Sonnenuntergang bis 10.Oktober 2024 Sonnenuntergang; Mittwoch Sonnenuntergang bis Donnerstag Sonnenuntergang]

6

6th Day | Yom Shishi | 10 October 2024 sunset to 11 October 2024 sunset; Thursday sunset until Friday sunset

[6.Tag: 10.Oktober 2024 Sonnenuntergang bis 11.Oktober 2024 Sonnenuntergang; Donnerstag Sonnenuntergang bis Freitag Sonnenuntergang]

7

7th Day | Shabbat | 11 October 2024 sunset to 12 October 2024 sunset; Friday sunset until Saturday sunset

[7.Tag: 11.Oktober 2024 Sonnenuntergang bis 12.Oktober 2024 Sonnenuntergang; Freitag Sonnenuntergang bis Samstag Sonnenuntergang]

8

1st Day | Yom Rishon | 12 October 2024 sunset to 13 October 2024 sunset; Saturday sunset until Sunday sunset

[1.Tag: 12.Oktober 2024 Sonnenuntergang bis 13.Oktober 2024 Sonnenuntergang; Samstag Sonnenuntergang bis Sonntag Sonnenuntergang]

9

2nd Day | Yom Sheni | 13 October 2024 sunset to 14 October 2024 sunset; Sunday sunset until Monday sunset

[2.Tag: 13.Oktober 2024 Sonnenuntergang bis 14.Oktober 2024 Sonnenuntergang; Sonntag Sonnenuntergang bis Montag Sonnenuntergang]

10

3rd Day | Yom Shlishi | 14 October 2024 sunset to 15 October 2024 sunset; Monday sunset until Tuesday sunset

[3.Tag: 14.Oktober 2024 Sonnenuntergang bis 15.Oktober 2024 Sonnenuntergang; Montag Sonnenuntergang bis Dienstag Sonnenuntergang]

11

4th Day | Yom Revi´i | 15 October 2024 sunset to 16 October 2024 sunset; Tuesday sunset until Wednesday sunset

[4.Tag: 15.Oktober 2024 Sonnenuntergang bis 16.Oktober 2024 Sonnenuntergang; Dienstag Sonnenuntergang bis Mittwoch Sonnenuntergang]

12

5th Day | Yom Hamishi | 16 October 2024 sunset to 17 October 2024 sunset; Wednesday sunset until Thursday sunset

[5.Tag: 16.Oktober 2024 Sonnenuntergang bis 17.Oktober 2024 Sonnenuntergang; Mittwoch Sonnenuntergang bis Donnerstag Sonnenuntergang]

13

6th Day | Yom Shishi | 17 October 2024 sunset to 18 October 2024 sunset; Thursday sunset until Friday sunset: Full Moon

[6.Tag: 17.Oktober 2024 Sonnenuntergang bis 18.Oktober 2024 Sonnenuntergang; Donnerstag Sonnenuntergang bis Freitag Sonnenuntergang: Vollmond]

14

7th Day | Shabbat | 18 October 2024 sunset to 19 October 2024 sunset; Friday sunset until Saturday sunset

[7.Tag: 18.Oktober 2024 Sonnenuntergang bis 19.Oktober 2024 Sonnenuntergang; Freitag Sonnenuntergang bis Samstag Sonnenuntergang]

15

1st Day | Yom Rishon | 19 October 2024 sunset to 20 October 2024 sunset; Saturday sunset until Sunday sunset

[1.Tag: 19.Oktober 2024 Sonnenuntergang bis 20.Oktober 2024 Sonnenuntergang; Samstag Sonnenuntergang bis Sonntag Sonnenuntergang]

16

2nd Day | Yom Sheni | 20 October 2024 sunset to 21 October 2024 sunset; Sunday sunset until Monday sunset

[2.Tag: 20.Oktober 2024 Sonnenuntergang bis 21.Oktober 2024 Sonnenuntergang; Sonntag Sonnenuntergang bis Montag Sonnenuntergang]

17

3rd Day | Yom Shlishi | 21 October 2024 sunset to 22 October 2024 sunset; Monday sunset until Tuesday sunset

[3.Tag: 21.Oktober 2024 Sonnenuntergang bis 22.Oktober 2024 Sonnenuntergang; Montag Sonnenuntergang bis Dienstag Sonnenuntergang]

18

4th Day | Yom Revi´i | 22 October 2024 sunset to 23 October 2024 sunset; Tuesday sunset until Wednesday sunset

[4.Tag: 22.Oktober 2024 Sonnenuntergang bis 23.Oktober 2024 Sonnenuntergang; Dienstag Sonnenuntergang bis Mittwoch Sonnenuntergang]

19

5th Day | Yom Hamishi | 23 October 2024 sunset to 24 October 2024 sunset; Wednesday sunset until Thursday sunset

[5.Tag: 23.Oktober 2024 Sonnenuntergang bis 24.Oktober 2024 Sonnenuntergang; Mittwoch Sonnenuntergang bis Donnerstag Sonnenuntergang]

20

6th Day | Yom Shishi | 24 October 2024 sunset to 25 October 2024 sunset; Thursday sunset until Friday sunset

[6.Tag: 24.Oktober 2024 Sonnenuntergang bis 25.Oktober 2024 Sonnenuntergang; Donnerstag Sonnenuntergang bis Freitag Sonnenuntergang]

21

7th Day | Shabbat | 25 October 2024 sunset to 26 October 2024 sunset; Friday sunset until Saturday sunset

[7.Tag: 25.Oktober 2024 Sonnenuntergang bis 26.Oktober 2024 Sonnenuntergang; Freitag Sonnenuntergang bis Samstag Sonnenuntergang]

22

1st Day | Yom Rishon | 26 October 2024 sunset to 27 October 2024 sunset; Saturday sunset until Sunday sunset

[1.Tag: 26.Oktober 2024 Sonnenuntergang bis 27.Oktober 2024 Sonnenuntergang; Samstag Sonnenuntergang bis Sonntag Sonnenuntergang]

23

2nd Day | Yom Sheni | 27 October 2024 sunset to 28 October 2024 sunset; Sunday sunset until Monday sunset

[2.Tag: 27.Oktober 2024 Sonnenuntergang bis 28.Oktober 2024 Sonnenuntergang; Sonntag Sonnenuntergang bis Montag Sonnenuntergang]

24

3rd Day | Yom Shlishi | 28 October 2024 sunset to 29 October 2024 sunset; Monday sunset until Tuesday sunset

[3.Tag: 28.Oktober 2024 Sonnenuntergang bis 29.Oktober 2024 Sonnenuntergang; Montag Sonnenuntergang bis Dienstag Sonnenuntergang]

25

4th Day | Yom Revi´i | 29 October 2024 sunset to 30 October 2024 sunset; Tuesday sunset until Wednesday sunset

[4.Tag: 29.Oktober 2024 Sonnenuntergang bis 30.Oktober 2024 Sonnenuntergang; Dienstag Sonnenuntergang bis Mittwoch Sonnenuntergang]

26

5th Day | Yom Hamishi | 30 October 2024 sunset to 31 October 2024 sunset; Wednesday sunset until Thursday sunset

[5.Tag: 30.Oktober 2024 Sonnenuntergang bis 31.Oktober 2024 Sonnenuntergang; Mittwoch Sonnenuntergang bis Donnerstag Sonnenuntergang]

27

6th Day | Yom Shishi | 31 October 2024 sunset to 1 November 2024 sunset; Thursday sunset until Friday sunset: Old Light

[6.Tag: 31.Oktober 2024 Sonnenuntergang bis 1.November 2024 Sonnenuntergang; Donnestag Sonnenuntergang bis Freitag Sonnenuntergang: Altlicht]

28

7th Day | Shabbat | 1 November 2024 sunset to 2 November 2024 sunset; Friday sunset until Saturday sunset: New Moon

[7.Tag: 1.November 2024 Sonnenuntergang bis 2.November 2024 Sonnenuntergang; Freitag Sonnenuntergang bis Samstag Sonnenuntergang: Neumond]

29 

1st Day | Yom Rishon | Rosh Chodesh: First Crescent Moon, sunset 3 Nov: New Light | 2 November 2024 sunset to 3 November 2024 sunset; Saturday sunset until Sunday sunset

[1.Tag: Rosh Chodesh: Erste sichtbare Mondsichel, Sonnenuntergang 3.Nov: Neulicht | 2.November 2024 Sonnenuntergang bis 3.November 2024 Sonnenuntergang; Samstag Sonnenuntergang bis Sonntag Sonnenuntergang]

30

Month 9 | *Monat 9*

Notes | Notizen

2nd Day | Yom Sheni | Rosh Chodesh: New Light | 3 November 2024 sunset to 4 November 2024 sunset; Sunday sunset until Monday sunset

[2.Tag: Rosh Chodesh: Neulicht | 3.November 2024 Sonnenuntergang bis 4.November 2024 Sonnenuntergang; Sonntag Sonnenuntergang bis Montag Sonnenuntergang]

1 

3rd Day | Yom Shlishi | 4 November 2024 sunset to 5 November 2024 sunset; Monday sunset until Tuesday sunset

[3.Tag: 4.November 2024 Sonnenuntergang bis 5.November 2024 Sonnenuntergang; Montag Sonnenuntergang bis Dienstag Sonnenuntergang]

2

4th Day | Yom Revi´i | 5 November 2024 sunset to 6 November 2024 sunset; Tuesday sunset until Wednesday sunset

[4.Tag: 5.November 2024 Sonnenuntergang bis 6.November 2024 Sonnenuntergang; Dienstag Sonnenuntergang bis Mittwoch Sonnenuntergang]

3

5th Day | Yom Hamishi | 6 November 2024 sunset to <u>7 November</u> 2024 sunset; Wednesday sunset until <u>Thursday</u> sunset

[5.Tag: 6.November 2024 Sonnenuntergang bis <u>7.November</u> 2024 Sonnenuntergang; Mittwoch Sonnenuntergang bis <u>Donnerstag</u> Sonnenuntergang]

6th Day | Yom Shishi | 7 November 2024 sunset to 8 November 2024 sunset; Thursday sunset until Friday sunset

[6.Tag: 7.November 2024 Sonnenuntergang bis 8.November 2024 Sonnenuntergang; Donnerstag Sonnenuntergang bis Freitag Sonnenuntergang]

5

7th Day | Shabbat | 8 November 2024 sunset to 9 November 2024 sunset; Friday sunset until Saturday sunset

[7.Tag: 8.November 2024 Sonnenuntergang bis 9.November 2024 Sonnenuntergang; Freitag Sonnenuntergang bis Samstag Sonnenuntergang]

6

1st Day | Yom Rishon | Death of Jannaeus: Golden Eagle Torn Down; Megillat Ta´anit | 9 November 2024 sunset to 10 November 2024 sunset; Saturday sunset until Sunday sunset

[1.Tag: Tod des Jannaeus: Niederriss des Goldenen Adlers; Megillat Ta´anit | 9.November 2024 Sonnenuntergang bis 10.November 2024 Sonnenuntergang; Samstag Sonnenuntergang bis Sonntag Sonnenuntergang]

7

2nd Day | Yom Sheni | Fast of 9th Month: Jehoiakim Burns Jeremiah´s Scroll (Jeremiah 36:1-32); Megillat Ta´anit | 10 November 2024 sunset to 11 November 2024 sunset; Sunday sunset until Monday sunset

[2.Tag: Fasten des 9.Monats: Jehoiakim verbrennt Jeremias Schriftrolle (Jeremia 36,1-32); Megillat Ta´anit | 10.November 2024 Sonnenuntergang bis 11.November 2024 Sonennuntergang; Sonntag Sonnenuntergang bis Montag Sonnenuntergang]

8

3rd Day | Yom Shlishi | 11 November 2024 sunset to 12 November 2024 sunset; Monday sunset until Tuesday sunset

[3.Tag: 11.November 2024 Sonnenuntergang bis 12.November 2024 Sonnenuntergang; Montag Sonnenuntergang bis Dienstag Sonnenuntergang]

9

4th Day | Yom Revi´i | 12 November 2024 sunset to 13 November 2024 sunset; Tuesday sunset until Wednesday sunset

[4.Tag: 12.November 2024 Sonnenuntergang bis 13.November 2024 Sonnenuntergang; Dienstag Sonnenuntergang bis Mittwoch Sonnenuntergang]

10

5th Day | Yom Hamishi | 13 November 2024 sunset to 14 November 2024 sunset; Wednesday sunset until Thursday sunset
[5.Tag: 13.November 2024 Sonnenuntergang bis 14.November 2024 Sonnenuntergang; Mittwoch Sonnenuntergang bis Donnerstag Sonnenuntergang]

11

6th Day | Yom Shishi | 14 November 2024 sunset to 15 November 2024 sunset; Thursday sunset until Friday sunset

[6.Tag: 14.November 2024 Sonnenuntergang bis 15.November 2024 Sonnenuntergang; Donnerstag Sonnenuntergang bis Freitag Sonnenuntergang]

12

7th Day | Shabbat | 15 November 2024 sunset to 16 November 2024 sunset; Friday sunset until Saturday sunset

[7.Tag: 15.November 2024 Sonnenuntergang bis 16.November 2024 Sonnenuntergang; Freitag Sonnenuntergang bis Samstag Sonnenuntergang]

13

1st Day | Yom Rishon | 16 November 2024 sunset to 17 November 2024 sunset; Saturday sunset until Sunday sunset: Full Moon

[1.Tag: 16.November 2024 Sonnenuntergang bis 17.November 2024 Sonnenuntergang; Samstag Sonnenuntergang bis Sonntag Sonnenuntergang: Vollmond]

14

2nd Day | Yom Sheni | 17 November 2024 sunset to 18 November 2024 sunset; Sunday sunset until Monday sunset

[2.Tag: 17.November 2024 Sonnenuntergang bis 18.November 2024 Sonnenuntergang; Sonntag Sonnenuntergang bis Montag Sonnenuntergang]

15

3rd Day | Yom Shlishi | 18 November 2024 sunset to 19 November 2024 sunset; Monday sunset until Tuesday sunset

[3.Tag: 18.November 2024 Sonnenuntergang bis 19.November 2024 Sonnenuntergang; Montag Sonnenuntergang bis Dienstag Sonnenuntergang]

16

4th Day | Yom Revi´i | 19 November 2024 sunset to 20 November 2024 sunset; Tuesday sunset until Wednesday sunset

[4.Tag: 19.November 2024 Sonnenuntergang bis 20.November 2024 Sonnenuntergang; Dienstag Sonnenuntergang bis Mittwoch Sonnenuntergang]

17

5th Day | Yom Hamishi | 20 November 2024 sunset to 21 November 2024 sunset; Wednesday sunset until Thursday sunset

[5.Tag: 20.November 2024 Sonnenuntergang bis 21.November 2024 Sonnenuntergang; Mittwoch Sonnenuntergang bis Donnerstag Sonnenuntergang]

18

6th Day | Yom Shishi | 21 November 2024 sunset to 22 November 2024 sunset; Thursday sunset until Friday sunset

[6.Tag: 21.November 2024 Sonnenuntergang bis 22.November 2024 Sonnenuntergang; Donnerstag Sonnenuntergang bis Freitag Sonnenuntergang]

19

7th Day | Shabbat | 22 November 2024 sunset to 23 November 2024 sunset; Friday sunset until Saturday sunset

[7.Tag: 22.November 2024 Sonnenuntergang bis 23.November 2024 Sonnenuntergang; Freitag Sonnenuntergang bis Samstag Sonnenuntergang]

20

1st Day | Yom Rishon | 23 November 2024 sunset to 24 November 2024 sunset; Saturday sunset until Sunday sunset

[1.Tag: 23.November 2024 Sonnenuntergang bis 24.November 2024 Sonnenuntergang; Samstag Sonnenuntergang bis Sonntag Sonnenuntergang]

21

2nd Day | Yom Sheni | 24 November 2024 sunset to 25 November 2024 sunset; Sunday sunset until Monday sunset

[2.Tag: 24.November 2024 Sonnenuntergang bis 25.November 2024 Sonnenuntergang; Sonntag Sonnenuntergang bis Montag Sonnenuntergang]

22

3rd Day | Yom Shlishi | 25 November 2024 sunset to 26 November 2024 sunset; Monday sunset until Tuesday sunset

[3.Tag: 25.November 2024 Sonnenuntergang bis 26.November 2024 Sonnenuntergang; Montag Sonnenuntergang bis Dienstag Sonnenuntergang]

23

4th Day | Yom Revi´i | 26 November 2024 sunset to 27 November 2024 sunset; Tuesday sunset until Wednesday sunset

[4.Tag: 26.November 2024 Sonnenuntergang bis 27.November 2024 Sonnenuntergang; Dienstag Sonnenuntergang bis Mittwoch Sonnenuntergang]

24

5th Day | Yom Hamishi | Hanukkah Day 1: Feast of Dedication (John 10:22) | 27 November 2024 sunset to 28 November 2024 sunset; Wednesday sunset until Thursday sunset

[5.Tag: Chanukka Tag 1: Einweihungsfest (Johannes 10,22) | 27.November 2024 Sonnenuntergang bis 28.November 2024 Sonnenuntergang; Mittwoch Sonnenuntergang bis Donnerstag Sonnenuntergang]

25

6th Day | Yom Shishi | Hanukkah Day 2: Feast of Dedication (John 10:22) | 28 November 2024 sunset to 29 November 2024 sunset; Thursday sunset until Friday sunset

[6.Tag: Chanukka Tag 2: Einweihungsfest (Johannes 10,22) | 28.November 2024 Sonnenuntergang bis 29.November 2024 Sonnenuntergang; Donnerstag Sonnenuntergang bis Freitag Sonnenuntergang]

26

7th Day | Shabbat | Hanukkah Day 3: Feast of Dedication (John 10:22) | Shabbat Hanukkah 1 | 29 November 2024 sunset to 30 November 2024 sunset; Friday sunset until Saturday sunset

[7.Tag: Shabbat Chanukka 1 | Hanukka Tag 3: Einweihungsfest (Johannes 10,22) | 29.November 2024 Sonnenuntergang bis 30.November 2024 Sonnenuntergang; Freitag Sonnenuntergang bis Samstag Sonnenuntergang]

27

1st Day | Yom Rishon | Hanukkah Day 4: Feast of Dedication (John 10:22) | 30 November 2024 sunset to 1 December 2024 sunset; Saturday sunset until Sunday sunset: Old Light

[1.Tag: Chanukka Tag 4: Einweihungsfest (Johannes 10,22) | 30.November 2024 Sonnenuntergang bis 1.Dezember 2024 Sonnenuntergang; Samstag Sonnenuntergang bis Sonntag Sonnenuntergang: Altlicht]

28 

2nd Day | Yom Sheni | Hanukkah Day 5: Feast of Dedication (John 10:22) | 1 December 2024 sunset to 2 December 2024 sunset; Sunday sunset until Monday sunset: New Moon

[2.Tag: Chanukka Tag 5: Einweihungsfest (Johannes 10,22) | 1.Dezember 2024 Sonnenuntergang bis 2.Dezember 2024 Sonnenuntergang; Sonntag Sonnenuntergang bis Montag Sonnenuntergang: Neumond]

29

3rd Day | Yom Shlishi | Rosh Chodes: First Crescent Moon, sunset 3 Dec: New Light | Hanukkah Day 6: Feast of Dedication (John 10:22) | 2 December 2024 sunset to 3 December 2024 sunset; Monday sunset until Tuesday sunset

[3.Tag: Rosh Chodesh: Erste sichtbare Mondsichel, Sonnenuntergang 3.Dez: Neulicht | Chanukka Tag 6: Einweihungsfest (Johannes 10,22) | 2.Dezember 2024 Sonnenunterang bis 3.Dezember 2024 Sonnenuntergang; Montag Sonnenuntergang bis Dienstag Sonnenuntergang]

30

Month 10 | *Monat 10*

Notes | Notizen

4th Day | Yom Revi´i | Rosh Chodesh: New Light | Hanukkah Day 7: Feast of Dedication (John 10:22) | 3 December 2024 sunset to 4 December 2024 sunset; Tuesday sunset until Wednesday sunset

[4.Tag: Rosh Chodesh: Neulicht | Chanukka Tag 7: Einweihungsfest (Johannes 10,22) | 3.Dezember 2024 Sonnenuntergang bis 4.Dezember 2024 Sonnenuntergang, Dienstag Sonnenuntergang bis Mittwoch Sonnenuntergang]

1

5th Day | Yom Hamishi | Hanukkah Day 8: Feast of Dedication (John 10:22) | 4 December 2024 sunset to 5 December 2024 sunset; Wednesday sunset until Thursday sunset

[5.Tag: Chanukka Tag 8: Einweihungsfest (Johannes 10,22) | 4.Dezember 2024 Sonnenuntergang bis 5.Dezember 2024 Sonnenuntergang; Mittwoch Sonnenuntergang bis Donnerstag Sonnenuntergang]

2

6th Day | Yom Shishi | 5 December 2024 sunset to 6 December 2024 sunset; Thursday sunset until Friday sunset

[6.Tag: 5.Dezember 2024 Sonnenuntergang bis 6.Dezember 2024 Sonnenuntergang; Donnerstag Sonnenuntergang bis Freitag Sonnenuntergang]

3

7th Day | Shabbat | 6 December 2024 sunset to 7 December 2024 sunset; Friday sunset to Saturday sunset

[7.Tag: 6.Dezember 2024 Sonnenuntergang bis 7.Dezember 2024 Sonnenuntergang; Freitag Sonnenuntergang bis Samstag Sonnenuntergang]

4

1st Day | Yom Rishon | 7 December 2024 sunset to 8 December 2024 sunset; Saturday sunset until Sunday sunset

[1.Tag: 7.Dezember 2024 Sonnenuntergang bis 8.Dezember 2024 Sonnenuntergang; Samstag Sonnenuntergang bis Sonntag Sonnenuntergang]

5

2nd Day | Yom Sheni | 8 December 2024 sunset to 9 December 2024 sunset; Sunday sunset until Monday sunset

[2.Tag: 8.Dezember 2024 Sonnenuntergang bis 9.Dezember 2024 Sonnenuntergang; Sonntag Sonnenuntergang bis Montag Sonnenuntergang]

6

3rd Day | Yom Shlishi | 9 December 2024 sunset to 10 December 2024 sunset; Monday sunset until Tuesday sunset

[3.Tag: 9.Dezember 2024 Sonnenuntergang bis 10.Dezember 2024 Sonnenuntergang; Montag Sonnenuntergang bis Dienstag Sonnenuntergang]

7

4th Day | Yom Revi´i | Festival of the Septuagint: Septuagint is Completed/Psalm 30:11; Megillat Ta´anit | 10 December 2024 sunset to 11 December 2024 sunset; Tuesday sunset until Wednesday sunset

[4.Tag: Fest der Septuaginta: Die Septuaginta ist vollendet/Psalm 30,11; Megillat Ta´anit | 10.Dezember 2024 Sonnenuntergang bid 11.Dezember 2024 Sonnenuntergang; Dienstag Sonnenuntergang bis Mittwoch Sonnenuntergang]

8

5th Day | Yom Hamishi | Unknown Fast: Ezra Dies/Selichot 10 Tevet; Ezkerah Mazok; Bahag, Kol Bo | 11 December 2024 sunset to 12 December 2024 sunset; Wednesday sunset until Thursday sunset

[5.Tag: Unbekanntes Fasten: Ezra stirbt/Selichot 10 Tevet; Ezkerah Mazok; Bahag, Kol Bo | 11.Dezember 2024 Sonnenuntergang bis 12.Dezember 2024 Sonnenuntergang; Mittwoch Sonnenuntergang bis Donnerstag Sonnenuntergang]

9

6th Day | Yom Shishi | Fast of Month 10: Siege of Jerusalem (2 Kings 25:1 | Jeremiah 52:4 | Ezekiel 24:1 | Zechariah 8:19) | 12 December 2024 sunset to 13 December 2024 sunset; Thursday sunset until Friday sunset

[6.Tag: Fasten des 10.Monats: Belagerung Jerusalems (2.Königs 25,1 | Jeremia 52,4 | Ezekiel 24,1 | Sacharja 8,19) | 12.Dezember 2024 Sonnenuntergang bis 13.Dezember 2024 Sonnenuntergang; Donnerstag Sonnenuntergang bis Freitag Sonnenuntergang]

10

7th Day | Shabbat | 13 December 2024 sunset to 14 December 2024 sunset; Friday sunset until Saturday sunset

[7.Tag: 13.Dezember 2024 Sonnenuntergang bis 14.Dezember 2024 Sonnenuntergang; Freitag Sonnenuntergang bis Samstag Sonnenuntergang]

11

1st Day | Yom Rishon | 14 December 2024 sunset to 15 December 2024 sunset; Saturday sunset until Sunday sunset

[1.Tag: 14.Dezember 2024 Sonnenuntergang bis 15.Dezember 2024 Sonnenuntergang; Samstag Sonnenuntergang bis Sonntag Sonnenuntergang]

12

2nd Day | Yom Sheni | 15 December 2024 sunset to 16 December 2024 sunset; Sunday sunset until Monday sunset

[2.Tag: 15.Dezember 2024 Sonnenuntergang bis 16.Dezember 2024 Sonnenuntergang; Sonntag Sonnenuntergang bis Montag Sonnenuntergang]

13

3rd Day | Yom Shlishi | 16 December 2024 sunset to 17 December 2024 sunset; Monday sunset until Tuesday sunset: Full Moon

[3.Tag: 16.Dezember 2024 Sonnenuntergang bis 17.Dezember 2024 Sonnenuntergang; Montag Sonnenuntergang bis Dienstag Sonnenuntergang:Vollmond]

14

4th Day | Yom Revi´i | 17 December 2024 sunset to 18 December 2024 sunset; Tuesday sunset until Wednesday sunset

[4.Tag: 17.Dezember 2024 Sonnenuntergang bis 18.Dezember 2024 Sonnenuntergang, Dienstag Sonnenuntergang bis Mittwoch Sonnenuntergang]

15

5th Day | Yom Hamishi | 18 December 2024 sunset to 19 December 2024 sunset; Wednesday sunset until Thursday sunset

[5.Tag: 18.Dezember 2024 Sonnenuntergang bis 19.Dezember 2024 Sonnenuntergang; Mittwoch Sonnenuntergang bis Donnerstag Sonnenuntergang]

16

6th Day | Yom Shishi | 19 December 2024 sunset to 20 December 2024 sunset; Thursday sunset until Friday sunset

[6.Tag: 19.Dezember 2024 Sonnenuntergang bis 20.Dezember 2024 Sonnenuntergang; Donnerstag Sonnenuntergang bis Freitag Sonnenuntergang]

17

7th Day | Shabbat | Winter Solstice (21 Dec, 09h:21m UT) | 20 December 2024 sunset to <u>21 December</u> 2024 sunset; Friday sunset until <u>Saturday</u> sunset

[7.Tag: Wintersonnenwende (21.Dez, 09h:21m UT) | 20.Dezember 2024 Sonnenuntergang bis <u>21.Dezember</u> 2024 Sonnenuntergang; Freitag Sonnenuntergang bis <u>Samstag</u> Sonnenuntergang]

18

1st Day | Yom Rishon | 21 December 2024 sunset to <u>22 December</u> 2024 sunset; Saturday sunset until <u>Sunday</u> sunset

[1.Tag: 21.Dezember 2024 Sonnenuntergang bis <u>22.Dezember</u> 2024 Sonnenuntergang; Samstag Sonnenuntergang bis <u>Sonntag</u> Sonnenuntergang]

19

2nd Day | Yom Sheni | 22 December 2024 sunset to 23 December 2024 sunset; Sunday sunset until Monday sunset

[2.Tag: 22.Dezember 2024 Sonnenuntergang bis 23.Dezember 2024 Sonnenuntergang; Sonntag Sonnenuntergang bis Montag Sonnenuntergang]

20

3rd Day | Yom Shlishi | 23 December 2024 sunset to 24 December 2024 sunset; Monday sunset until Tuesday sunset

[3.Tag: 23.Dezember 2024 Sonnenuntergang bis 24.Dezember 2024 Sonnenuntergang; Montag Sonnenuntergang bis Dienstag Sonnenuntergang]

21

4th Day | Yom Revi´i | 24 December 2024 sunset to 25 December 2024 sunset; Tuesday sunset until Wednesday sunset

[4.Tag: 24.Dezember 2024 Sonnenuntergang bis 25.Dezember 2024 Sonnenuntergang; Dienstag Sonnenuntergang bis Mittwoch Sonnenuntergang]

22

5th Day | Yom Hamishi | 25 December 2024 sunset to 26 December 2024 sunset; Wednesday sunset until Thursday sunset

[5.Tag: 25.Dezember 2024 Sonnenuntergang bis 26.Dezember 2024 Sonnenuntergang; Mittwoch Sonnenuntergang bis Donnerstag Sonnenuntergang]

23

6th Day | Yom Shishi | 26 December 2024 sunset to 27 December 2024 sunset; Thursday sunset until Friday sunset

[6.Tag: 26.Dezember 2024 Sonnenuntergang bis 27.Dezember 2024 Sonnenuntergang; Donnerstag Sonnenuntergang bis Freitag Sonnenuntergang]

24

7th Day | Shabbat | 27 December 2024 sunset to 28 December 2024 sunset; Friday sunset until Saturday sunset

[7.Tag: 27.Dezember 2024 Sonnenuntergang bis 28.Dezember 2024 Sonnenuntergang, Freitag Sonnenuntergang bis Samstag Sonnenuntergang]

25

1st Day | Yom Rishon | 28 December 2024 sunset to 29 December 2024 sunset; Saturday sunset until Sunday sunset

[1.Tag: 28.Dezember 2024 Sonnenuntergang bis 29.Dezember 2024 Sonnenuntergang; Samstag Sonnenuntergang bis Sonntag Sonnenuntergang]

26

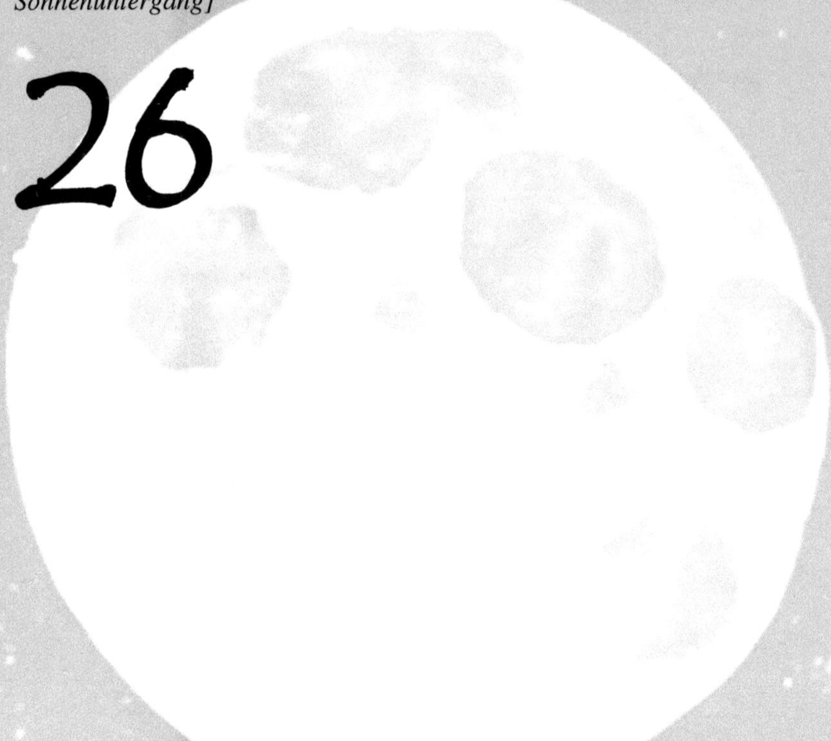

2nd Day | Yom Sheni | 29 December 2024 sunset to 30 December 2024 sunset; Sunday sunset until Monday sunset: Old Light

[2.Tag: 29.Dezember 2024 Sonnenuntergang bis 30.Dezember 2024 Sonnenuntergang; Sonntag Sonnenuntergang bis Montag Sonnenuntergang: Altlicht]

27

3rd Day | Yom Shlishi | 30 December 2024 sunset to 31 December 2024 sunset; Monday sunset until Tuesday sunset: New Moon

[3.Tag: 30.Dezember 2024 Sonnenuntergang bis 31.Dezember 2024 Sonnenuntergang; Montag Sonnenuntergang bis Dienstag Sonnenuntergang: Neumond]

28

4th Day | Yom Revi´i | Rosh Chodesh: First Crescent Moon, sunset 1 Jan: New Light | 31 December 2024 sunset to 1 January 2025 sunset; Tuesday sunset until Wednesday sunset

[4.Tag: Rosh Chodesh: Erste sichtbare Mondsichel, Sonnenuntergang 1.Jan: Neulicht | 31.Dezember 2024 Sonnenuntergang bis 1.Januar 2025 Sonnenuntergang; Dienstag Sonnenuntergang bis Mittwoch Sonnenuntergang]

29 

Notes | Notizen

Month 11 | *Monat 11*

Notes | Notizen

5th Day | Yom Hamishi | Rosh Chodesh: First Crescent Moon, sunset 1 Jan: New Light | 1 January 2025 sunset to <u>2 January</u> 2025 sunset; Wednesday sunset until <u>Thursday</u> sunset

[5.Tag: Rosh Chodesh: Erste sichtbare Mondsichel, Sonnenuntergang 1.Jan: Neulicht | 1.Januar 2025 Sonnenuntergang bis <u>2.Januar</u> 2025 Sonnenuntergang; Mittwoch Sonnenuntergang bis <u>Donnerstag</u> Sonnenuntergang]

1 

6th Day | Yom Shishi | Death of Herod: Matthew 2:19; Megillat Ta´anit | 2 Janauary 2025 sunset to <u>3 January</u> 2025 sunset; Thursday sunset until <u>Friday</u> sunset

[6.Tag: Tod von Herodes: Matthäus 2,19; Megillat Ta´anit | 2.Januar 2025 Sonnenuntergang bis <u>3.Januar</u> 2025 Sonnenuntergang; Donnerstag Sonnenuntergang bis <u>Freitag</u> Sonnenuntergang]

2

7th Day | Shabbat | 3 January 2025 sunset to 4 January 2025 sunset; Friday sunset until Saturday sunset

[7.Tag: 3.Januar 2025 Sonnenuntergang bis 4.Januar 2025 Sonnenuntergang; Freitag Sonnenuntergang bis Samstag Sonnenuntergang]

3

1st Day | Yom Rishon | 4 January 2025 sunset to 5 January 2025 sunset; Saturday sunset until Sunday sunset

[1.Tag: 4.Januar 2025 Sonnenuntergang bis 5.Januar 2025 Sonnenuntergang; Samstag Sonnenuntergang bis Sonntag Sonnenuntergang]

4

2nd Day | Yom Sheni | 5 January 2025 sunset to 6 January 2025 sunset; Sunday sunset until Monday sunset

[2.Tag: 5.Januar 2025 Sonnenuntergang bis 6.Januar 2025 Sonnenuntergang; Sonntag Sonnenuntergang bis Montag Sonnenuntergang]

5

3rd Day | Yom Shlishi | 6 January 2025 sunset to 7 January 2025 sunset; Monday sunset until Tuesday sunset

[3.Tag: 6.Januar 2025 Sonnenuntergang bis 7.Januar 2025 Sonnenuntergang; Montag Sonnenuntergang bis Dienstag Sonnenuntergang]

6

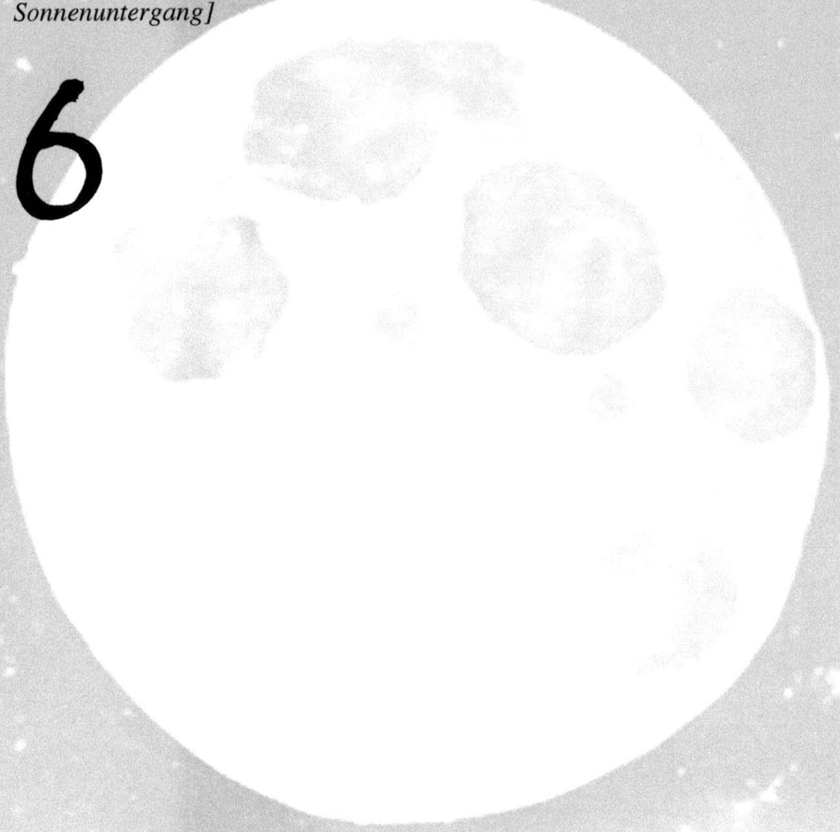

4th Day | Yom Revi´i | 7 January 2025 sunset to 8 January 2025 sunset; Tuesday sunset until Wednesday sunset

[4.Tag: 7.Januar 2025 Sonnenuntergang bis 8.Januar 2025 Sonnenuntergang; Dienstag Sonnenuntergang bis Mittwoch Sonnenuntergang]

7

5th Day | Yom Hamishi | 8 January 2025 sunset to 9 January 2025 sunset; Wednesday sunset until Thursday sunset

[5.Tag: 8.Januar 2025 Sonnenuntergang bis 9.Januar 2025 Sonnenuntergang; Mittwoch Sonnenuntergang bis Donnerstag Sonnenuntergang]

8

6th Day | Yom Shishi | 9 January 2025 sunset to 10 January 2025 sunset; Thursday sunset until Friday sunset

[6.Tag: 9.Januar 2025 Sonnenuntergang bis 10.Januar 2025 Sonnenuntergang; Donnerstag Sonnenuntergang bis Freitag Sonnenuntergang]

9

7th Day | Shabbat | 10 January 2025 sunset to 11 January 2025 sunset; Friday sunset until Saturday sunset

[7.Tag: 10.Januar 2025 Sonnenuntergang bis 11.Januar 2025 Sonnenuntergang; Freitag Sonnenuntergang bis Samstag Sonnenuntergang]

10

1st Day | Yom Rishon | 11 January 2025 sunset to 12 January 2025 sunset; Saturday sunset until Sunday sunset

[1.Tag: 11.Januar 2025 Sonnenuntergang bis 12.Januar 2025 Sonnenuntergang; Samstag Sonnenuntergang bis Sonntag Sonnenuntergang]

11

2nd Day | Yom Sheni | 12 January 2025 sunset to 13 January 2025 sunset; Sunday sunset until Monday sunset

[2.Tag: 12.Januar 2025 Sonnenuntergang bis 13.Januar 2025 Sonnenuntergang; Sonntag Sonnenuntergang bis Montag Sonnenuntergang]

12

3rd Day | Yom Shlishi | 13 January 2025 sunset to 14 January 2025 sunset; Monday sunset until Tuesday sunset

[3.Tag: 13.Januar 2025 Sonnenuntergang bis 14.Januar 2025 Sonnenuntergang; Montag Sonnenuntergang bis Dienstag Sonnenuntergang]

13

4th Day | Yom Revi´i | 14 January 2025 sunset to 15 January 2025 sunset; Tuesday sunset until Wednesday sunset: Full Moon

[4.Tag: 14.Januar 2025 Sonnenuntergang bis 15.Januar 2025 Sonnenuntergang; Dienstag Sonnenuntergang bis Mittwoch Sonnenuntergang: Vollmond]

14

5th Day | Yom Hamishi | ***Tu B´Shvat: New Year for Trees*****/Rosh Hashanah 14A & B | 15 January 2025 sunset to 16 January 2025 sunset; Wednesday sunset until Thursday sunset**

[5.Tag: Tu B´Shvat: Neujahr der Bäume/*Rosh Hashana 14A & B | 15.Januar 2025 Sonnenuntergang bis 16.Januar 2025 Sonnenuntergang; Mittwoch Sonnenuntergang bis Donnerstag Sonnenuntergang]*

15

6th Day | Yom Shishi | 16 January 2025 sunset to 17 January 2025 sunset; Thursday sunset until Friday sunset

[6.Tag: 16.Januar 2025 Sonnenuntergang bis 17.Januar 2025 Sonnenuntergang; Donnerstag Sonnenuntergang bis Freitag Sonnenuntergang]

16

7th Day | Shabbat | 17 January 2025 sunset to 18 January 2025 sunset; Friday sunset until Saturday sunset

[7.Tag: 17.Januar 2025 Sonnenuntergang bis 18.Januar 2025 Sonnenuntergang; Freitag Sonnenuntergang bis Samstag Sonnenuntergang]

17

1st Day | Yom Rishon | 18 January 2025 sunset to 19 January 2025 sunset; Saturday sunset until Sunday sunset

[1.Tag: 18.Januar 2025 Sonnenuntergang bis 19.Januar 2025 Sonnenuntergang; Samstag Sonnenuntergang bis Sonntag Sonnenuntergang]

18

2nd Day | Yom Sheni | 19 January 2025 sunset to 20 January 2025 sunset; Sunday sunset until Monday sunset

[2.Tag: 19.Januar 2025 Sonnenuntergang bis 20.Januar 2025 Sonnenuntergang; Sonntag Sonnenuntergang bis Montag Sonnenuntergang]

19

3rd Day | Yom Shlishi | 20 January 2025 sunset to 21 January 2025 sunset; Monday sunset until Tuesday sunset

[3.Tag: 20.Januar 2025 Sonnenuntergang bis 21.Januar 2025 Sonnenuntergang; Montag Sonnenuntergang bis Dienstag Sonnenuntergang]

20

4th Day | Yom Revi´i | 21 January 2025 sunset to 22 January 2025 sunset, Tuesday sunset until Wednesday sunset

[4.Tag: 21.Januar 2025 Sonnenuntergang bis 22.Januar 2025 Sonnenuntergang; Dienstag Sonnenuntergang bis Mittwoch Sonnenuntergang]

21

5th Day | Yom Hamishi | 22 January 2025 sunset to 23 January 2025 sunset; Wednesday sunset until Thursday sunset

[5.Tag: 22.Januar 2025 Sonnenuntergang bis 23.Januar 2025 Sonnenuntergang; Mittwoch Sonnenuntergang bis Donnerstag Sonnenuntergang]

22

6th Day | Yom Shishi | 23 January 2025 sunset to 24 January 2025 sunset; Thursday sunset until Friday sunset

[6.Tag: 23.Januar 2025 Sonnenuntergang bis 24.Januar 2025 Sonnenuntergang; Donnerstag Sonnenuntergang bis Freitag Sonnenuntergang]

23

7th Day | Shabbat | 24 January 2025 sunset to 25 January 2025 sunset; Friday sunset until Saturday sunset

[7.Tag: 24.Januar 2025 Sonnenuntergang bis 25.Januar 2025 Sonnenuntergang; Freitag Sonnenuntergang bis Samstag Sonnenuntergang]

24

1st Day | Yom Rishon | 25 January 2025 sunset to 26 January 2025 sunset; Saturday sunset until Sunday sunset

[1.Tag: 25.Januar 2025 Sonnenuntergang bis 26.Januar 2025 Sonnenuntergang; Samstag Sonnenuntergang bis Sonntag Sonnenuntergang]

25

2nd Day | Yom Sheni | 26 January 2025 sunset to 27 January 2025 sunset; Sunday sunset until Monday sunset

[2.Tag: 26.Januar 2025 Sonnenuntergang bis 27.Januar 2025 Sonnenuntergang; Sonntag Sonnenuntergang bis Montag Sonnenuntergang]

26

3rd Day | Yom Shlishi | 27 January 2025 sunset to 28 January 2025 sunset; Monday sunset until Tuesday sunset: Old Light

[3.Tag: 27.Januar 2025 Sonnenuntergang bis 28.Januar 2025 Sonnenuntergang: Montag Sonnenuntergang bis Dienstag Sonnenuntergang: Altlicht]

27

4th Day | Yom Revi´i | 28 January 2025 sunset to 29 January 2025 sunset; Tuesday sunset until Wednesday sunset: New Moon

[4.Tag: 28.Januar 2025 Sonnenuntergang bis 29.Januar 2025 Sonnenuntergang; Dienstag Sonnenuntergang bis Mittwoch Sonnenuntergang: Neumond]

28

5th Day | Yom Hamishi | Rosh Chodesh: First Crescent Moon, sunset 30 Jan: New Light | 29 January 2025 sunset to 30 January 2025 sunset; Wednesday sunset until Thursday sunset

[5.Tag: Rosh Chodesh: Erste sichtbare Mondsichel, Sonnenuntergang 30.Jan: Neulicht | 29.Januar 2025 Sonnenuntergang bis 30.Januar 2025 Sonnenuntergang; Mittwoch Sonnenuntergang bis Donnerstag Sonnenuntergang]

29 

Notes | Notizen

Month 12 | *Monat 12*

Notes | Notizen

6th Day | Yom Shishi | Rosh Chodesh: New Light | 30 January 2025 sunset to 31 January 2025 sunset; Thursday sunset until Friday sunset

[6.Tag: Rosh Chodesh: Neulicht | 30.Januar 2025 Sonnenuntergang bis 31.Januar 2025 Sonnenuntergang; Donnerstag Sonnenuntergang bis Freitag Sonnenuntergang]

1 

7th Day | Shabbat | 31 January 2025 sunset to 1 February 2025 sunset; Friday sunset until Saturday sunset

[7.Tag: 31.Januar 2025 Sonnenuntergang bis 1.Februar 2025 Sonnenuntergang; Freitag Sonnenuntergang bis Samstag Sonnenuntergang]

2

1st Day | Yom Rishon | 1 February 2025 sunset to 2 February 2025 sunset; Saturday sunset until Sunday sunset

[1.Tag: 1.Februar 2025 Sonnenuntergang bis 2.Februar 2025 Sonnenuntergang; Samstag Sonnenuntergang bis Sonntag Sonnenuntergang]

3

2nd Day | Yom Sheni | 2 February 2025 sunset to 3 February 2025 sunset; Sunday sunset until Monday sunset

[2.Tag: 2.Februar 2025 Sonnenuntergang bis 3.Februar 2025 Sonnenuntergang; Sonntag Sonnenuntergang bis Montag Sonnenuntergang]

3rd Day | Yom Shlishi | 3 February 2025 sunset to 4 February 2025 sunset; Monday sunset until Tuesday sunset

[3.Tag: 3.Februar 2025 Sonnenuntergang bis 4.Februar 2025 Sonnenuntergang; Montag Sonnenuntergang bis Dienstag Sonnenuntergang]

5

4th Day | Yom Revi´i | 4 February 2025 sunset to 5 February sunset; Tuesday sunset until Wednesday sunset

[4.Tag: 4.Februar 2025 Sonnenuntergang bis 5.Februar 2025 Sonnenuntergang; Dienstag Sonnenuntergang bis Mittwoch Sonnenuntergang]

6

5th Day | Yom Hamishi | 5 February 2025 sunset to 6 February 2025 sunset; Wednesday sunset until Thursday sunset

[5.Tag: 5.Februar 2025 Sonnenuntergang bis 6.Februar 2025 Sonnenuntergang; Mittwoch Sonnenuntergang bis Donnerstag Sonnenuntergang]

7

6th Day | Yom Shishi | 6 February 2025 sunset to 7 February 2025 sunset; Thursday sunset until Friday sunset

[6.Tag: 6.Februar 2025 Sonnenuntergang bis 7.Februar 2025 Sonnenuntergang; Donnerstag Sonnenuntergang bis Freitag Sonnenuntergang]

8

7th Day | Shabbat | 7 February 2025 sunset to 8 February 2025 sunset; Friday sunset until Saturday sunset

[7.Tag: 7.Februar 2025 Sonnenuntergang bis 8.Februar 2025 Sonnenuntergang; Freitag Sonnenuntergang bis Samstag Sonnenuntergang]

9

1st Day | Yom Rishon | 8 February 2025 sunset to <u>9 February</u> 2025 sunset; Saturday sunset until <u>Sunday</u> sunset

[1.Tag: 8.Februar 2025 Sonnenuntergang bis <u>9.Februar</u> 2025 Sonnenuntergang; Samstag Sonnenuntergang bis <u>Sonntag</u> Sonnenuntergang]

10

2nd Day | Yom Sheni | 9 February 2025 sunset to 10 February 2025 sunset; Sunday sunset until Monday sunset

[2.Tag: 9.Februar 2025 Sonnenuntergang bis 10.Februar 2025 Sonnenuntergang; Sonntag Sonnenuntergang bis Montag Sonnenuntergang]

11

3rd Day | Yom Shlishi | 10 February 2025 sunset to 11 February 2025 sunset; Monday sunset until Tuesday sunset

[3.Tag: 10.Februar 2025 Sonnenuntergang bis 11.Februar 2025 Sonnenuntergang; Montag Sonnenuntergang bis Dienstag Sonnenuntergang]

12

4th Day | Yom Revi´i | 11 February 2025 sunset to 12 February 2025 sunset; Tuesday sunset until Wednesday sunset

[4.Tag: 11.Februar 2025 Sonnenuntergang bis 12.Februar 2025 Sonnenuntergang; Dienstag Sonnenuntergang bis Mittwoch Sonnenuntergang]

13

5th Day | Yom Hamishi | 12 February 2025 sunset to 13 February 2025 sunset; Wednesday sunset until Thursday sunset: Full Moon

[5.Tag: 12.Februar 2025 Sonnenuntergang bis 13.Februar 2025 Sonnenuntergang; Mittwoch Sonnenuntergang bis Donnerstag Sonnenuntergang: Vollmond]

14

6th Day | Yom Shishi | 13 February 2025 sunset to 14 February 2025 sunset; Thursday sunset until Friday sunset

[6.Tag: 13.Februar 2025 Sonnenuntergang bis 14.Februar 2025 Sonnenuntergang; Donnerstag Sonnenuntergang bis Freitag Sonnenuntergang]

15

7th Day | Shabbat | 14 February 2025 sunset to 15 February 2025 sunset; Friday sunset until Saturday sunset

[7.Tag: 14.Februar 2025 Sonnenuntergang bis 15.Februar 2025 Sonnenuntergang; Freitag Sonnenuntergang bis Samstag Sonnenuntergang]

16

1st Day | Yom Rishon | 15 February 2025 sunset to <u>16 February</u> 2025 sunset; Saturday sunset until <u>Sunday</u> sunset

[1.Tag: 15.Februar 2025 Sonnenuntergang bis <u>16.Februar</u> 2025 Sonnenuntergang; Samstag Sonnenuntergang bis <u>Sonntag</u> Sonnenuntergang]

17

2nd Day | Yom Sheni | 16 February 2025 sunset to 17 February 2025 sunset; Sunday sunset until Monday sunset

[2.Tag: 16.Februar 2025 Sonnenuntergang bis 17.Februar 2025 Sonnenuntergang; Sonntag Sonnenuntergang bis Montag Sonnenuntergang]

18

3rd Day | Yom Shlishi | 17 February 2025 sunset to 18 February 2025 sunset; Monday sunset until Tuesday sunset

[3.Tag: 17.Februar 2025 Sonnenuntergang bis 18.Februar 2025 Sonnenuntergang; Montag Sonnenuntergang bis Dienstag Sonnenuntergang]

19

4th Day | Yom Revi´i | 18 February 2025 sunset to 19 February 2025 sunset; Tuesday sunset until Wednesday sunset

[4.Tag: 18.Februar 2025 Sonnenuntergang bis 19.Februar 2025 Sonnenuntergang; Dienstag Sonnenuntergang bis Mittwoch Sonnenuntergang]

20

5th Day | Yom Hamishi | 19 February 2025 sunset to 20 February 2025 sunset; Wednesday sunset until Thursday sunset

[5.Tag: 19.Februar 2025 Sonnenuntergang bis 20.Februar 2025 Sonnenuntergang; Mittwoch Sonnenuntergang bis Donnerstag Sonnenuntergang]

21

6th Day | Yom Shishi | 20 February 2025 sunset to 21 February 2025 sunset; Thursday sunset until Friday sunset

[6.Tag: 20.Februar 2025 Sonnenuntergang bis 21.Februar 2025 Sonnenuntergang, Donnerstag Sonnenuntergang bis Freitag Sonnenuntergang]

22

7th Day | Shabbat | 21 February 2025 sunset to 22 February 2025 sunset; Friday sunset until Saturday sunset

[7.Tag: 21.Februar 2025 Sonnenuntergang bis 22.Februar 2025 Sonnenuntergang; Freitag Sonnenuntergang bis Samstag Sonnenuntergang]

23

1st Day | Yom Rishon | 22 February 2025 sunset to <u>23 February</u> 2025 sunset; Saturday sunset until <u>Sunday</u> sunset

[1.Tag: 22.Februar 2025 Sonnenuntergang bis <u>23.Februar</u> 2025 Sonnenuntergang; Samstag Sonnenuntergang bis <u>Sonntag</u> Sonnenuntergang]

24

2nd Day | Yom Sheni | 23 February 2025 sunset to 24 February 2025 sunset; Sunday sunset until Monday sunset

[2.Tag: 23.Februar 2025 Sonnenuntergang bis 24.Februar 2025 Sonnenuntergang; Sonntag Sonnenuntergang bis Montag Sonnenuntergang]

25

3rd Day | Yom Shlishi | 24 February 2025 sunset to 25 February sunset; Monday sunset until Tuesday sunset

[3.Tag: 24.Februar 2025 Sonnenuntergang bis 25.Februar 2025 Sonnenuntergang; Montag Sonnenuntergang bis Dienstag Sonnenuntergang]

26

4th Day | Yom Revi´i | 25 February 2025 sunset to <u>26 February</u> 2025 sunset; Tuesday sunset until <u>Wednesday</u> sunset

[4.Tag: 25.Februar 2025 Sonnenuntergang bis <u>26.Februar</u> 2025 Sonnenuntergang; Dienstag Sonnenuntergang bis <u>Mittwoch</u> Sonnenuntergang]

27

5th Day | Yom Hamishi | 26 February 2025 sunset to 27 February 2025 sunset; Wednesday sunset until Thursday sunset: Old Light

[5.Tag: 26.Februar 2025 Sonnenuntergang bis 27.Februar 2025 Sonnenuntergang; Mittwoch Sonnenuntergang bis Donnestag Sonnenuntergang; Altlicht]

28

6th Day | Yom Shishi | 27 February 2025 sunset to 28 February 2025 sunset; Thursday sunset until Friday sunset: New Moon

[6.Tag: 27: Februar 2025 Sonnenuntergang bis 28.Februar 2025 Sonnenuntergang; Donnerstag Sonnenuntergang bis Freitag Sonnenuntergang: Neumond]

29

7th Day | Shabbat | Rosh Chodesh: First Crescent Moon, sunset 1 March: New Light | 28 February 2025 sunset to 1 March 2025 sunset; Friday sunset until Saturday sunset

[7.Tag: Rosh Chodesh: Erste sichtbare Mondsichel, Sonnenuntergang 1.März: Neulicht | 28.Februar 2025 Sonnenuntergang bis 1.März 2025 Sonnenuntergang; Freitag Sonnenuntergang bis Samstag Sonnenuntergang]

30 

Month 13 | *Monat 13*

Notes | Notizen

1st Day | Yom Rishon | Rosh Chodesh: New Light | 1 March 2025 sunset to 2 March 2025 sunset; Saturday sunset until Sunday sunset

[1.Tag: Rosh Chodesh: Neulicht | 1.März 2025 Sonnenuntergang bis 2.März 2025 Sonnenuntergang; Samstag Sonnenuntergang bis Sonntag Sonnenuntergang]

1 

2nd Day | Yom Sheni | 2 March 2025 sunset to 3 March 2025 sunset; Sunday sunset until Monday sunset

[2.Tag: 2.März 2025 Sonnenuntergang bis 3.März 2025 Sonnenuntergang; Sonntag Sonnenuntergang bis Montag Sonnenuntergang]

2

3rd Day | Yom Shlishi | 3 March 2025 sunset to 4 March 2025 sunset; Monday sunset until Tuesday sunset

[3.Tag: 3.März 2025 Sonnenuntergang bis 4.März 2025 Sonnenuntergang; Montag Sonnenuntergang bis Dienstag Sonnenuntergang]

3

4th Day | Yom Revi´i | 4 March 2025 sunset to 5 March 2025 sunset; Tuesday sunset until Wednesday sunset

[4.Tag: 4.März 2025 Sonnenuntergang bis 5.März 2025 Sonnenuntergang; Dienstag Sonnenuntergang bis Mittwoch Sonnenuntergang]

5th Day | Yom Hamishi | 5 March 2025 sunset to 6 March 2025 sunset; Wednesday sunset until Thursday sunset

[5.Tag: 5.März 2025 Sonnenuntergang bis 6.März 2025 Sonnenuntergang; Mittwoch Sonnenuntergang bis Donnerstag Sonnenuntergang]

5

6th Day | Yom Shishi | 6 March 2025 sunset to 7 March 2025 sunset; Thursday sunset until Friday sunset

[6.Tag: 6.März 2025 Sonnenuntergang bis 7.März 2025 Sonnenuntergang; Donnerstag Sonnenuntergang bis Freitag Sonnenuntergang]

6

7th Day | Shabbat | 7 March 2025 sunset to 8 March 2025 sunset; Friday sunset until Saturday sunset

[7.Tag: 7.März 2025 Sonnenuntergang bis 8.März 2025 Sonnenuntergang; Freitag Sonnenuntergang bis Samstag Sonnenuntergang]

7

1st Day | Yom Rishon | 8 March 2025 sunset to 9 March 2025 sunset; Saturday sunset until Sunday sunset

[1.Tag: 8.März 2025 Sonnenuntergang bis 9.März 2025 Sonnenuntergang; Samstag Sonnenuntergang bis Sonntag Sonnenuntergang]

8

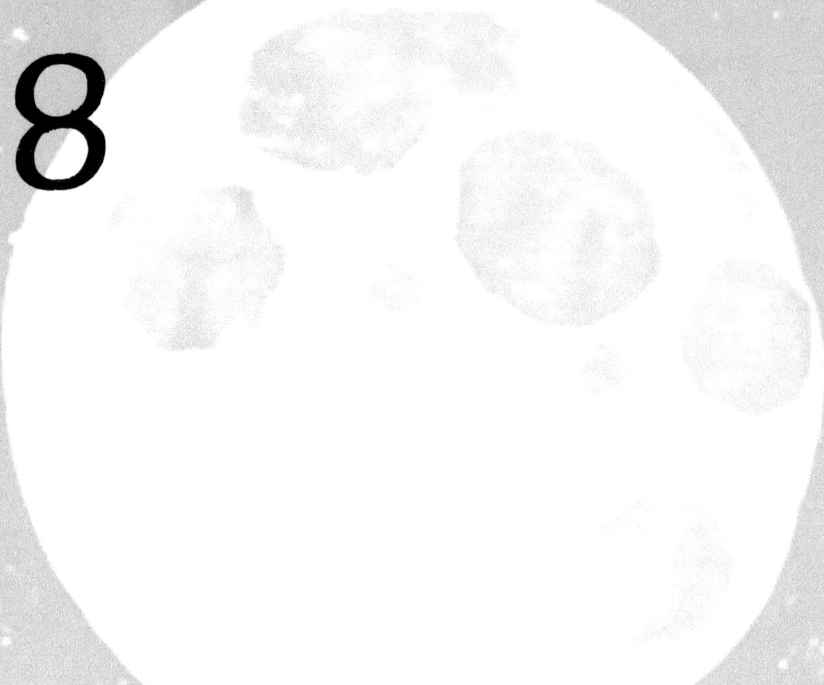

2nd Day | Yom Sheni | 9 March 2025 sunset to 10 March 2025 sunset; Sunday sunset until Monday sunset

[2.Tag: 9.März 2025 Sonnenuntergang bis 10.März 2025 Sonnenuntergang; Sonntag Sonnenuntergang bis Montag Sonnenuntergang]

9

3rd Day | Yom Shlishi | 10 March 2025 sunset to 11 March 2025 sunset; Monday sunset until Tuesday sunset

[3.Tag: 10.März 2025 Sonnenuntergang bis 11.März 2025 Sonnenuntergang; Montag Sonnenuntergang bis Dienstag Sonnenuntergang]

10

4th Day | Yom Revi´i | 11 March 2025 sunset to 12 March 2025 sunset; Tuesday sunset until Wednesday sunset

[4.Tag: 11.März 2025 Sonnenuntergang bis 12.März 2025 Sonnenuntergang; Dienstag Sonnenuntergang bis Mittwoch Sonnenuntergang]

11

5th Day | Yom Hamishi | 12 March 2025 sunset to 13 March 2025 sunset; Wednesday sunset until Thursday sunset

[5.Tag: 12.März 2025 Sonnenuntergang bis 13.März 2025 Sonnenuntergang; Mittwoch Sonnenuntergang bis Donnerstag Sonnenuntergang]

12

6th Day | Yom Shishi | Total Lunar Eclipse | Ta´anit Esther: Commemorative Fast of Esther [Esther 9:1-14; Megillat 2A] | 13 March 2025 sunset to 14 March 2025 sunset; Thursday sunset until Friday sunset

[6.Tag: Totale Mondfinsternis | Ta´anit Ester: Ester-Gedenkfasten [Ester 9,1-14; Megillat 2A] | 13.März 2025 Sonnenuntergang bis 14.März 2025 Sonnenuntergang; Donnerstag Sonnenuntergang bis Freitag Sonnenuntergang]

13

7th Day | Shabbat | Purim Lots (Esther 9:15-17) | 14 March 2025 sunset to 15 March 2025 sunset; Friday sunset until Saturday sunset: Full Moon

[7.Tag: Purim Lose (Ester 9,15-17) | 14.März 2025 Sonnenuntergang bis 15.März 2025 Sonnenuntergang; Freitag Sonnenuntergang bis Samstag Sonnenuntergang: Vollmond]

14

1st Day | Yom Rishon | Shushan Purim Lots (Esther 9:18-32) | 15 March 2025 sunset to 16 March 2025 sunset; Saturday sunset until Sunday sunset

[1.Tag: Shushan Purim Lose/Purim der Städet (Ester 9,18-32) | 15.März 2025 Sonnenuntergang bis 16.März 2025 Sonnenuntergang; Samstag Sonnenuntergang bis Sonntag Sonnenuntergang]

15

2nd Day | Yom Sheni | 16 March 2025 sunset to 17 March 2025 sunset; Sunday sunset until Monday sunset

[2.Tag: 16.März 2025 Sonnenuntergang bis 17.März 2025 Sonnenuntergang; Sonntag Sonnenuntergang bis Montag Sonnenuntergang]

16

3rd Day | Yom Shlishi | 17 March 2025 sunset to <u>18 March</u> 2025 sunset; Monday sunset until <u>Tuesday</u> sunset

[3.Tag: 17.März 2025 Sonnenuntergang bis <u>18.März</u> 2025 Sonnenuntergang; Montag Sonnenuntergang bis <u>Dienstag</u> Sonnenuntergang]

17

4th Day | Yom Revi´i | 18 March 2025 sunset to 19 March 2025 sunset; Tuesday sunset until Wednesday sunset

[4.Tag: 18.März 2025 Sonnenuntergang bis 19.März 2025 Sonnenuntergang; Dienstag Sonnenuntergang bis Mittwoch Sonnenuntergang]

18

5th Day | Yom Hamishi | Spring Equinox (20 Mar, 09h:02m UT) | 19 March 2025 sunset to 20 March 2025 sunset; Wednesday sunset until Thursday sunset

[5.Tag: Frühlingsäquinoktium (20.März, 09h:02m UT) | 19.März 2025 Sonnenuntergang bis 20.März 2025 Sonnenuntergang; Mittwoch Sonnenuntergang bis Donnerstag Sonnenuntergang]

19

6th Day | Yom Shishi | 20 March 2025 sunset to 21 March 2025 sunset; Thursday sunset until Friday sunset

[6.Tag: 20.März 2025 Sonnenuntergang bis 21.März 2025 Sonnenuntergang; Donnerstag Sonnenuntergang bis Freitag Sonnenuntergang]

20

7th Day | Shabbat | 21 March 2025 sunset to 22 March 2025 sunset; Friday sunset until Saturday sunset

[7.Tag: 21.März 2025 Sonnenuntergang bis 22.März 2025 Sonnenuntergang; Freitag Sonnenuntergang bis Samstag Sonnenuntergang]

21

1st Day | Yom Rishon | 22 March 2025 sunset to 23 March 2025 sunset; Saturday sunset until Sunday sunset

[1.Tag: 22.März 2025 Sonnenuntergang bis 23.März 2025 Sonnenuntergang; Samstag Sonnenuntergang bis Sonntag Sonnenuntergang]

22

2nd Day | Yom Sheni | 23 March 2025 sunset to 24 March 2025 sunset; Sunday sunset until Monday sunset

[2.Tag: 23.März 2025 Sonnenuntergang bis 24.März 2025 Sonnenuntergang; Sonntag Sonnenuntergang bis Montag Sonnenuntergang]

23

3rd Day | Yom Shlishi | 24 March 2025 sunset to 25 March 2025 sunset; Monday sunset until Tuesday sunset

[3.Tag: 24.März 2025 Sonnenuntergang bis 25.März 2025 Sonnenuntergang; Montag Sonnenuntergang bis Dienstag Sonnenuntergang]

24

4th Day | Yom Revi´i | 25 March 2025 sunset to 26 March 2025 sunset; Tuesday sunset until Wednesday sunset

[4.Tag: 25.März 2025 Sonnenuntergang bis 26.März 2025 Sonnenuntergang; Dienstag Sonnenuntergang bis Mittwoch Sonnenuntergang]

25

5th Day | Yom Hamishi | 26 March 2025 sunset to 27 March 2025 sunset; Wednesday sunset until Thursday sunset

[5.Tag: 26.März 2025 Sonnenuntergang bis 27.März 2025 Sonnenuntergang; Mittwoch Sonnenuntergang bis Donnerstag Sonnenuntergang]

26

6th Day | Yom Shishi | 27 March 2025 sunset to 28 March 2025 sunset; Thursday sunset until Friday sunset: Old Light

[6.Tag: 27.März 2025 Sonnenuntergang bis 28.März 2025 Sonnenuntergang; Donnerstag Sonnenuntergang bis Freitag Sonnenuntergang: Altlicht]

27

7th Day | Shabbat | 28 March 2025 sunset to 29 March 2025 sunset; Friday sunset until Saturday sunset: New Moon & Partial Solar Eclipse

[7.Tag: 28.März 2025 Sonnenuntergang bis 29.März 2025 Sonnenuntergang; Freitag Sonnenuntergang bis Samstag Sonnenuntergang: Neumond & Partielle Sonnenfinsternis]

28  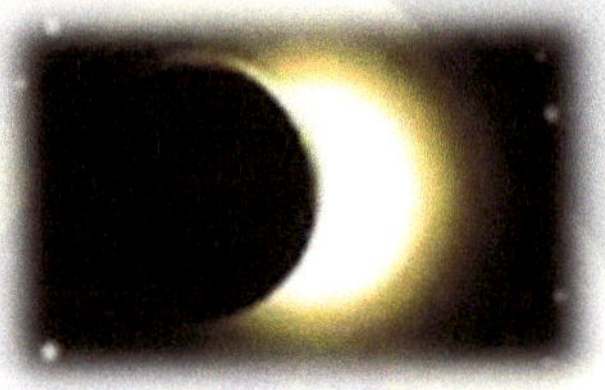

1st Day | Yom Rishon | Rosh Chodesh: First Crescent Moon, sunset 30 Mar: New Light | 29 March 2025 sunset to 30 March 2025 sunset; Saturday sunset until Sunday sunset

[1.Tag: Rosh Chodesh: Erste sichtbare Mondsichel, Sonnenuntergang 30.März: Neulicht | 29.März 2025 Sonnenuntergang bis 30.März 2025 Sonnenuntergang; Samstag Sonnenuntergang bis Sonntag Sonnenuntergang]

29 

Notes | Notizen

Month 1 | *Monat 1*

Abib

Barley ripening (ripening of ears)

Gerstenreife (Ährenreife)

2nd Day | Yom Sheni | 30 March 2025 sunset to 31 March 2025 sunset; Sunday sunset until Monday sunset |
[2.Tag: 30.März 2025 Sonnenuntergang bis 31.März 2025 Sonnenuntergang; Sonntag Sonnenuntergang bis Montag Sonnenuntergang]

Rosh Chodesh: First Visible Crescent Moon seen at sunset on 30 March 2025

[Erste sichtbare Mondsichel zu sehen am 30.März 2025 bei Sonnenuntergang]

Rosh Chodeshim:

Spiritual Official New Year (Exodus 12:1-2) by Elohim | *Geistiges Offizielles Neujahr (Exodus 12,1-2) durch Elohim*

Notes | Notizen

526

529